Subjectivities in Investigation of *the Urban*

The Scream, the Shadow and the Mirror

都市の探求における主観性：叫びと影と鏡像

|ıılı|lıı|lıı|ı|ıı|ıl| Measuring the Non-Measurable 06

Acknowledgments

Subjectivities in Investigation of the Urban: The Scream, the Shadow and the Mirror was produced within the major research project *Measuring the Non-Measurable*, conducted by International Keio Institute for Architecture and Urbanism – IKI (2011–13) and supported by the Japanese Ministry of Education and Training Strategic Research Grant.

Subjectivities in Investigation of the Urban: The Scream, the Shadow and the Mirror presents parts of *Mn'M* explorations of urban quality and the results of the fieldwork conducted by the *Mn'M* research team in Hong Kong (24–26 May, 2013), Bangkok (26–29 June, 2013) and Singapore (3–6 July, 2013). It brings together individual and collective efforts of the core *Mn'M* Asian case study members and research assistants. As the Head of the *Mn'M* research project, I wish to give my sincere thanks to all participants of the Asia *dérive* sessions and contributors to visual essays which constitute a significant part of this volume: Chinese University of Hong Kong team (Hendrik Tieben, Thomas Chung, Mika Savela, Mo Kar Him, Cherry Wong, Della Cheung, Daniel Ho, Bill So); Chulalongkorn and Silpakorn University, Bangkok teams (Apiradee Kasemsook, Nuttinee Karnchanaporn, Wimonrart Issarathumnoon, Monsicha Kanittaprasert, Thiti Kunajitpimol, Methi Laithavewat, Prat Tinrat, Parima Amnuaywattana, Dherapat Sanguandekul, Siwat Wittayalertanya, Kanyapat Seneewong Na Ayudhaya, Sasinee Teangtawat, Thittayabhorn Mitudom), National University of Singapore team (Heng Chye Kiang, Yeo Su-Jan, Patricia Chia, Lim Szeying) and Keio University co+labo Radović Tokyo team (Davisi Boontharm, Tamao Hashimoto, Milica Muminović, Ilze Paklone, Ken Akatsuka and Yoshiaki Kato), for their commitment, dedication, generous contribution and collegial atmosphere throughout all stages of production of this volume. I also give my heartfelt thanks to Boris Kuk, an outstanding graphic designer from Croatia, whose creative inputs and discussions inspired and helped the editorial team reach higher communicative quality of visual essays in this *volume*.

The material presented in this book was produced during, and in the weeks that followed *dérive* sessions held in Hong Kong, Bangkok and Singapore. The fieldwork in three cities would not be possible without the support of numerous academics and students from the School of Architecture, Chinese University of Hong Kong, Faculty of Architecture, Chulalongkorn University, Faculty of Architecture, Silpakorn University, and School of Design and Environment, National University of Singapore, to which all *dérive* members are wholeheartedly grateful.

Lastly, my very special thanks go to the *Mn'M* visual essays editorial team: Milica Muminović, Ilze Paklone, Tamao Hashimoto and Rafael A. Balboa who selflessly contributed their time. energy and creative efforts to consolidate and reinterpret the heterogeneous fieldwork material; to driving forces behind the production of this volume, Shinya Takagi and Takako Ishida (flick studio), Kyushu Pistol graphical design team (Tomofumi Yoshida and Yu Nakao), and production coordinator Kumi Aizawa (silent voice). Special thanks also goes to the Japanese language translator Masako Hirobe and Masako Okamoto (TAC Co., Ltd.).

Last, but not the least, the team behind production of this volume expresses gratitude to Keio University, International Keio Institute for Architecture and Urbanism – IKI and co+labo Radović for providing supportive conditions which made this volume possible.

Darko Radović

Subjectivities in Investigation of *the Urban* : The Scream, the Shadow and the Mirror
都市の探求における主観性〜叫びと影と鏡像

A PROLOGUE	**Measuring the non-Measurable – the Search of Urban Qualities Which Reaches Beyond the Conventional** Measuring the Non-Measurable —— 定石を超えて、都市の質を探求する Darko Radović｜ダルコ・ラドヴィッチ	004
INTRODUCTION	**Thinking About Cities – As If People Mattered**　都市を思考する——人間を中心として Darko Radović｜ダルコ・ラドヴィッチ	006
ESSAYS	**On Subjectivities in Investigation of *the Urban*** 都市研究における主観性について Darko Radović｜ダルコ・ラドヴィッチ	008
	The Scream, the Shadow and the Mirror 叫びと影と鏡像 Darko Radović｜ダルコ・ラドヴィッチ	022
	These Visual Essays ... ヴィジュアルエッセイについて… Darko Radović｜ダルコ・ラドヴィッチ	024
ASIA DÉRIVE	**HONG KONG DÉRIVE** 香港漂流 Visual Essay by Tamao Hashimoto, Ilze Paklone and Rafael A. Balboa ヴィジュアルエッセイ：橋本圭央、イルゼ・パクロネ、ラファエル・バルボア	026
	BANGKOK DÉRIVE バンコク漂流 Visual Essay by Milica Muminović ヴィジュアルエッセイ：ミリッツァ・ムミノヴィッチ	050
	SINGAPORE DÉRIVE シンガポール漂流 Visual Essay by Ilze Paklone and Rafael A. Balboa ヴィジュアルエッセイ：イルゼ・パクロネ、ラファエル・バルボア	082
THE EPILOGUE	Darko Radović｜ダルコ・ラドヴィッチ	113

A PROLOGUE
Measuring the Non-Measurable — The Search of Urban Qualities which Reaches Beyond the Conventional

Measuring the Non-Measurable──定石を超えて、都市の質を探求する

Darko Radović | ダルコ・ラドヴィッチ

The primary focus of the research project *Measuring the non-Measurable – Mn'M*, within which the research for this book has been conducted, is at two systems of urban phenomena which resist (easy or any) quantification – culture and sustainability. *Mn'M* enters the field of debates about urban quality by challenging the very idea of measurability. "Measuring" in *Mn'M* is, thus, just a shorthand for all efforts to capture and represent quality solely in a "scientific" way. Over the years, the research team members from ten cities in Asia, Europe and Australia joined forces to challenge the unsustainable schism between "measurable" and "non-measurable", "textual" and "numerical" in *the* urban and address the totality, the Lefebvrian oeuvre.

Despite the best and the most honest of efforts, all attempts to express the key qualities of urban spaces and practices fall short of capturing their defining feature – an untamable complexity. Standard tools of urban research and analysis tend to be reductionist, favouring some aspects of *the urban* over the other, either to fit some pre-defined brief or, simply, due to the very impossibility to measure quality which makes cities – cities. In order to meet the challenge, at broadest, conceptual level, *Mn'M* investigates relationships between system theory (social sciences and engineering) and assemblage theory (Deleuze, De Landa, Dovey). Its cross-disciplinary theoretical frameworks rely on place theory, Lefebvrian social theories and the idea of *eco-urbanity* (Radović). At methodological levels, the project combines standard research practices with emerging sensibilities and some resurgent critical practices, such as Situationist subversions and psychogeography (Debord).

This book is mainly about the final of those three methodological attempts, as it forcefully theorises and explains the necessity to address complex urban phenomena with correspondingly complex methods. It proposes an inclusion of some aspects of human experience *the urban* which are commonly proscribed in research – such as subjective insights and sensuality, as an explicit recognition of the multisensorial nature of our being in the world.

Subjectivities in Investigation of the Urban: The Scream, the Shadow and the Mirror is about searching, finding (or, at least, strongly believing that one has found) and communicating the subtlest aspects of urban quality, those which tend to escape easy definition. It presents parts of the work conducted for *Mn'M* research project within co+labo Radović, an architecture and urban design laboratory of Keio University, Tokyo, and its associate academic partners, members of the *Mn'M* core team from Hong Kong, Bangkok and Singapore. The work was orchestrated around intensive fieldwork sessions inspired by the Situationists practice of *dérive*. The focus of these methodologically radical experiments is at recognizing, identifying, variously capturing and sharing diverse spatial expressions of urban intensity *without shutting-up feelings, intuition and sensuality of the participants*. That contributes to better thinking about the phenomena which elude efforts to be captured and quantified, while, at the same time, making critical contribution to the *oeuvre*, the ultimate quality of *the urban*.

This book theorises that approach by directly addressing its most contentions aspects – inclusion of subjectivity of the researcher her/himself, and elusive dimensions of multisensory reality of our interaction with the environment. It opens by addressing the key issues of relevance for investigation of qualitative dimensions urban spaces and practices, and then further explores the key themes through immersion in specific locations in the cities of Hong Kong, Bangkok and Singapore. That is where, through a number of intensive *dérive* sessions, interactions between the researchers of different individual sensibilities and cultural backgrounds, and those concrete urban places, unfolded, and where the material for three visual essays has been compiled. Dialectisations of various degrees of foreignness and local familiarities opened those three, palpably distinctive urban cultures to unorthodox research insights, the selection of which is presented in this book. The dual emphasis is on (1) layered subjectivities and knotted multi-sensorial nuances of places and practices which the teams and the team members have encountered, and (2) the challenge of (re)presentation and communication of those experiences (in the case of the book – to the reader. The book is, thus, about alternative knowledges, appreciations, and representations of such knowledges and of *the urban*.

Asia *dérive* experiment was conducted in the period May–July 2013, in collaboration between Keio University (co+labo radović) and research teams from the Chinese University of Hong Kong (headed by Hendrik Tieben and Thomas Chung), Chulalongkorn University (Wimonrart Issarathumnoon) and Silpakorn University, Bangkok (Apiradee Kasemsook and Nuttinee Karnchanaporn), and National University of Singapore (Heng Chye Kiang).

DR, February 2014

「Measuring the non-Measurable (Mn'M)」研究プロジェクトは主に、都市の現象のうち(安易に、あるいはどのような形でも)数量化されることを拒む2種類の体系──文化と持続可能性に重点を置いている。Mn'Mは測定可能という概念そのものに挑戦することで、都市の質の議論に踏み込むものであり、したがって、Mn'Mにおける「Measuring＝測定する」という言葉は、「科学的な」方法だけで質を捉えて説明しようとするあらゆる試みを簡潔に言い表したものに過ぎない。Mn'Mプロジェクトそのものは、都市の重要な質と、その複雑さに取り組むさまざまな方法について議論する場として誕生した。アジア、ヨーロッパ、オーストラリアの10都市から集まったメンバーを中心として研究チームが構成され、「都市的なるもの」の「測定可能」な側面と「測定不可能」な側面、「感触的」側面と「数量的」側面が無理に分離されていることに異議を呈し、あるいはルフェーブル的「作品」としての全体に取り組めるところにまで到達することを目指している。

都市空間と都市活動については、これまでその本質を表現しようと最大限のまっとうな努力が払われてきたにもかかわらず、未だどの試みもその決定的な特徴──飼い慣らすことのできない複雑性を捉えるには至っていない。都市研究や都市分析のための標準的なツールは、論旨を正当化するためか、あるいは単に、都市を都市たらしめている質を測定することがまさしく不可能であるという理由からか、「都市的なるもの」のある側面を他の側面よりも重視した、還元主義的なものになりがちである。この課題に取り組むために、Mn'Mは極めて広範に概念的レベルでシステム論(社会科学、社会工学)と集合論(Deleuze, De Landa, Dovey)の関係を探究する。その理論的枠組みは学際的であり、場所論、ルフェーブル的社会概念、「エコ・アーバニティ」概念(Radović)に依拠している。方法論レベルでは、感情の発現に関する標準的調査法と、シチュアシオニストが提唱した転覆と心理地理学(Debord)といった重要な方法を再構築したものを組み合わせている。

本書では、複雑な現象に複雑な手法で取り組むことの必要性についても併せて議論する。ここでは個人、主観的見識、そして人間とは世界を五感で捉えるものであるという明確な認識など、通常の研究では排除されるような、人間による「都市的なるもの」の体験という観点も含めることを提唱している。

『Subjectivities in Investigations of the Urban: The Scream, the Shadow and the Mirror (都市の探求における主観性：叫びと影と鏡像)』は、都市の質のなかでも極めて捉えにくく、普通はたやすく定義することのできない側面を探求し、発見(あるいは少なくとも発見したと確信)し、伝えようとするものである。掲載されているのは、東京の慶応義塾大学で建築都市デザインに取り組むラドヴィッチ研究室(co+labo)が、香港、バンコク、シンガポールからMn'M研究チームの中心メンバーとして集まった学術パートナーと共に実施したMn'Mの研究内容の一部である。この研究では、シチュアシオニストが行った「漂流」のアイデアを取り入れつつ、さまざまなフィールドワークが行われた。こうした現地実験の第一弾は2012年秋に東京で実施され、その成果の一部が本シリーズの『東京漂流──都市の強度を探して』として出版されている。

この実験では、「感触や直感、場所感覚を閉ざすことなく」、ラディカルな方法論を用いて、多様な空間的発現としての都市強度を認識し、特定し、さまざまな形で捉え、共有することに主眼が置かれている。そうしたかたちで実験を行うことで、捉えられて数量化されることを拒みながらも、同時に「作品」──「都市的なるもの」の質に大きくかかわっているような現象について、より深く考察することができるのである。

本書では、この手法のなかで最も論議の的になりやすい点として、特に研究者自身の主観を持ち込むことと、我々がまわりの環境を五感で体験しているという捉えどころのない現実的側面を扱ったうえで、この手法について論じていく。まず都市空間と都市活動の質的側面の調査に関して重要な事項を扱い、次にそれらのテーマを香港、バンコク、シンガポール各都市の特定の場所に適用して検証する。そこでは度重なる「漂流」や、それぞれ感覚も文化的背景も異なる研究者らによる実在都市空間との積極的な交流が行われるわけであるが、そのなかで集められたものが材料となって3つのヴィジュアル・エッセイが構成されている。さまざまなレベルの異質性と現地に対する親密性を弁証法化することによって、明白な特徴をもつこれら3つの都市文化を研究対象として深く考察する道が開かれたものであり、そのなかから選りすぐられた内容が本書に掲載されている。ここで重視されているのは、研究チームのメンバーが出会った場所や活動に対する多重的な主観、それらを五感で捉えたニュアンス、そしてどのようにそれを読者に提示して伝えるかという方法である。したがって、本書は「都市的なるもの」についての、もうひとつの知見と見解であり、その表現なのである。

アジア「漂流」実験は2013年5月から7月にかけて、慶應義塾大学ラドヴィッチ研究室(co+labo)が、香港中文大学(研究チームリーダー：ヘンドリック・ティーベン、トーマス・チュン)、バンコクのチュラーロンコーン大学(ウィモンラット・イッサラータッマヌーン)とシラパコーン大学(アピラディ・カセンスーク、ナッティニ・カルンチャポン)、シンガポール国立大学(王才强)の研究チームと共同で実施したものである。

INTRODUCTION
Thinking about Cities
— As if People Mattered

都市を思考する──人間を中心として

Darko Radović | ダルコ・ラドヴィッチ

This book puts forward one simple, but contentions issue, namely the need for non-reductive approach to investigations of *the urban*. It is deliberately polemological, in hope to, in the tradition of Michel de Certeau, "force theory to recognise its own limits" (Highmore, 2006).

While urban analysis seeks objectivity, in reality that quality often stays elusive. Although many aspects of making and living cities critically rely on both natural and social sciences, the complexity of the urban is never reducible to the scientific thinking only. Reaching the very core of human condition, the existential nature of cities often defies analytical rigour and logic. Nevertheless, immense urban complexity needs careful planning, responsible design and sensitive management. That demands constant dialectisation of various knowledges (Haraway, 1991). In order to make itself useful, much of the thinking has to be simplified and pragmatically instrumentalised, so that it can address the real goals within the real, strict time-frames. The success of many necessary, pragmatic actions critically relies upon reliable, objective, scientifically accurate methods and databases, the best that rigorous urban research, planning and design-research are capable of providing. That is true for all sorts of dimensions of urban functioning, which need responsible planning, design, development and management, to ensure optimal performance of complex urban systems.

The problem arises when that same logic gets applied to the fragile and lively socio-cultural fabric which makes the very essence of *the urban*. In simplest terms, when applying techno-scientific logic to Lefebvrian *oeuvre*, we easily end up with practices which could dwarf the darkest of Orwellian fears. The paradigms which support such thinking indeed tend to be totalitarian, and thus genuinely unable to comprehend that there are the spheres of human and urban reality where it would be not only unnecessary, but where it is directly counterproductive and dangerous to be simply "efficient" and "useful". Within the world of booming techno-optimism, that question needs to be urgently addressed, and addressing it makes the core of our pledge for non-reductive approach to *the urban*. We need flexible and robust, composite approaches capable of addressing, with equal relevance, competence and confidence, both *measurable* and *non-measurable*, *numerical* and *textual* (Radović, 2012, 2013, 2013a) dimensions of urban phenomena and variety of their dynamic interactions. Communicating the resulting knowledges is as important as those knowledges themselves, for the ultimate aim is to empower the bottom-up energies and support recognition of countless, layered subjective realities – all of which matter.

Another dimension of this issue should not be forgotten. The ruling paradigm is not based only on benevolent, but misplaced technocratic desire for efficiency. As hinted above, there is also a strong undercurrent of totalitarian political will which stimulates such management, design, planning practices and, most dangerously, there is a corresponding reductivist and instrumentalised thinking about *the urban*. The *desired* outcome of such thinking and its application is precisely the reduction of *the urban*, in response to an explicit intention to tame the innate, potentially explosive social intensities which only dense, complex urban environments can generate. A bewildering fascination with innovation, and seemingly unlimited support for creating the (brave) new (world) and, in particular, for producing highly controllable, parallel, virtual and addictive realities transforms the agora into the panopticon – at the global scale.

Without further dwelling on the socio-political aspects of academic discourse (which we have covered in the previous books of *Mn'M* edition), here we only summarise that: (1) there is an evident support for research which emphasizes measurable, quantifiable aspects of urban quality; (2) such trends come from the fact that global power brokers worship a single bottom line; and that (3) urban research, seeking legitimization, credibility and funds from such sources of power, all too often presents itself as capable of identifying and establishing the unshakeable facts, to measure and to "prove". Their results better reflect the realities and demands of the dominant power than actual socio-cultural conditions on the ground. In return, the simplified and instrumentalised worldviews which they support simplify and instrumentalise the world. That was never as obvious as it is today, when the mere 0.01% of world population, literally the richest 85 individuals across the globe, share a combined wealth equal to the possessions of the poorest 3.5 billion of the world's population (Oxfam, 2014). As the gap between the elites and population, the majority of which is urban, widens dramatically, one side arguably receives some very measurable benefits, while the vast majority ends up with the glaringly non-measurable misery. That is not accidental. Globalisation, as we know it today, was designed to efficiently concentrate power and wealth, while the multiplicity and diversity of human realities remain bewilderingly complex and hopelessly powerless. Urban research plays active role in producing such realities.

The main focus of this book is at the very point in which the need for a well-founded, all-inclusive, non-reductionist thinking about *the urban* clashes with efficiency of the reductionist approaches of contemporary, "spectacular" (Debord, 1998) urbanism. The aim is to point at some dimensions of the urban which get routinely pushed aside, by bringing them (back) into the practices of thinking, feeling, planning, designing, making, managing and *living cities* – as if people mattered (Schumacher, 1973).

本書は、単純ながらも議論の分かれるひとつのテーマ、すなわち「都市的なるもの」の探求に対する非還元主義的アプローチの必要性を提唱するものである。ミシェル・ド・セルトーの「理論に理論そのものの限界を認識させる」(Highmore, 2006) という流儀を受け継げればとの思いから、あえて戦争論的な展開となっている。

都市の分析には客観性が求められるが、実際のところ都市の質は往々にして捉えどころがないままである。都市づくりや都市生活は多くの点で自然科学と社会科学の両方に依存してはいるものの、「都市的なるもの」の複雑性は決して科学的思考のみに還元できるものではない。人間のありようの核心に迫るならば、分析の厳密性も論理性も、大抵は都市の実存的性質によって拒まれるのである。それでもやはり、とてつもない複雑性を孕む都市というものは、周到に計画され、責任をもってデザインされ、慎重に管理されることを必要としている。これには、さまざまな知識を絶え間なく弁証法化することが求められる (Haraway, 1991)。思考することも構築することも、それ自体が有用なものとなるために、大部分は単純化され実際に道具として利用されなくてはならず、そうすることによって現実の厳しい時間枠の中で実際の目標に対応することができるのである。必要とされている数々の実際的行為が成功するかどうかは、厳密な都市研究、都市計画や都市デザインの研究が得意とする、確実で客観的かつ科学的に正確な手法とデータベースにかかっている。これは都市の機能のあらゆる側面について言えることであり、複雑な都市体系が最適なかたちで機能するためには、責任ある計画、デザイン、開発、管理が必要なのである。

問題なのは、「都市的なるもの」のまさしく本質をつくり上げている脆くもあり躍動的でもある社会文化的構造に、これと同じ論理があてはめられたときである。簡単な言い方をすれば、科学技術的な論理をルフェーブル的「作品」に適用すると、いとも簡単に、オーウェリアン的暗黒世界に対する恐怖が陳腐に感じられる状況に陥ってしまうのだ。このような思考の背後にある理論枠組みは全体主義的傾向があり、そしてまた、人間と都市の現実には、不必要であるばかりか直接的には非生産的で、「効率的」や「便利」であることが危険でもある領域が存在するということが、この枠組みの中では理解できない。技術楽観論が幅を利かせる世界において、これは早急に取り組まなくてはならない問題であり、この問題に取り組むことこそ、「都市的なるもの」に対して非還元主義的アプローチを行うという我々の誓いの核心をなすものである。我々に必要なのは、都市の現象と、その多様でダイナミックな相互作用について、「測定可能」な側面と「測定不可能」な側面、「数量的」側面と「感触的」側面 (Radović, 2012, 2013) に、いずれも等しい妥当性と力量と確信をもって取り組むことができるような、柔軟で確固たる複合的アプローチである。その結果として得られた知見を伝えることは、知見そのものと同じぐらい重要だ。究極の目標は、ボトムアップのエネルギーを活性化させ、幾重にも重なる主観的現実の認識を促進することであり、それこそが大切なのだから。

この問題にはもうひとつ、忘れてはならない側面がある。この支配的な理論的枠組みは、いくばくかの善意にだけではなく、効率を求めるテクノクラートの筋違いな欲望にも根ざしているということである。そして上記で示唆したように、その奥底には強力で全体主義的な政治的意思も存在しており、そのような管理、デザイン、計画を推進するばかりか、さらに極めて危険なことに、それに呼応した還元主義的で道具主義的な思考で「都市的なるもの」を考えることを助長するのである。彼らにとって「望ましい結果」とは、「都市的なるもの」の還元に他ならない。濃密で複雑な都市環境だけがつくり出すことのできる社会強度は本来的に爆発でもしかねないようなものであるが、この社会強度を飼い慣らすことを明確に意図しているのである。「すばらしい新世界」の創造、つまり極めて管理的でパラレルかつヴァーチャルな中毒性の現実をつくり出すことを際限なく支援し、やみくもに技術革新に陶酔することは、世界規模でアゴラ (広場) をパノプティコン (全展望監視システム) に変えることになる。

ここでは、Mn'Mシリーズの既刊本 (Radović 2013) で取り上げたアカデミック・ディスコースの社会政治的側面に深入りはせず、おおまかに概要を述べるにとどめておくが、第一に言えることは、都市の質のなかでも測定と数量化の可能な側面ばかりを重視した研究が支持を得ていることは明らかである。第二に、この風潮は、世界的な権力を握る実力者らが、経済的側面だけを重視するシングルボトムライン崇拝の世界を描いているという事実から生まれている。そして第三に、都市研究そのものが、そのような権力の源からの認知と信用と資金とを求めるあまり、ゆるぎない事実の特定と確立、さらには測定と「証明」の可能なものとして自己をアピールしているケースが多すぎる。それ故に、研究手法は還元主義的なものになりがちである。研究結果は、実際に現場で発生している社会文化的状況よりも、支配権力層の現実と欲求を色濃く反映したものとなる。それと引き換えに、権力層が支持する単純化された道具主義的な世界観によって、世界は単純化され道具化されるのである。こうした状況が今ほどあからさまな時代はない。世界人口のわずか0.01％に過ぎない大富豪85人が、全世界の貧困層35億人の財産総額に相当する富を握っているのである (Oxfam, 2014)。エリートと一般住民 (大半が都市に居住) との格差は劇的に広がっており、前者がかなりの測定可能な利益を間違いなく手にしている一方で、圧倒的多数は明らかに測定不可能な貧困に陥っている。これは偶然ではない。現在のグローバリゼーションは富と権力を効率的に集約するようにできており、人間的現実の多様性は途方に暮れるほど複雑でどうしようもなく無力なままだというのに、都市研究は積極的に今のような現実をつくり出す役目を果たしているのである。

十分な根拠に基づいて包括的かつ非還元主義的に「都市的なるもの」を思考することの必要性は、現代の「スペクタクルな」(Debord, 1998) アーバニズムの還元主義的アプローチがもつ効率性とぶつかり合う。まさにその部分が、本書で最も重きを置いている点である。本書は、「都市的なるもの」がもつ側面のなかで隅に押しやられている部分を示そうとするものであり、そのために、都市を思考し、感じ、計画し、デザインし、構築し、管理し、都市に「暮らす」という活動の中に、そうした側面をもち込む (取り戻す) ことを試みるものである——人間を中心 (Schumacher, 1973) として。

On Subjectivities in Investigation of *the Urban*

都市研究における主観性について

Darko Radović | ダルコ・ラドヴィッチ

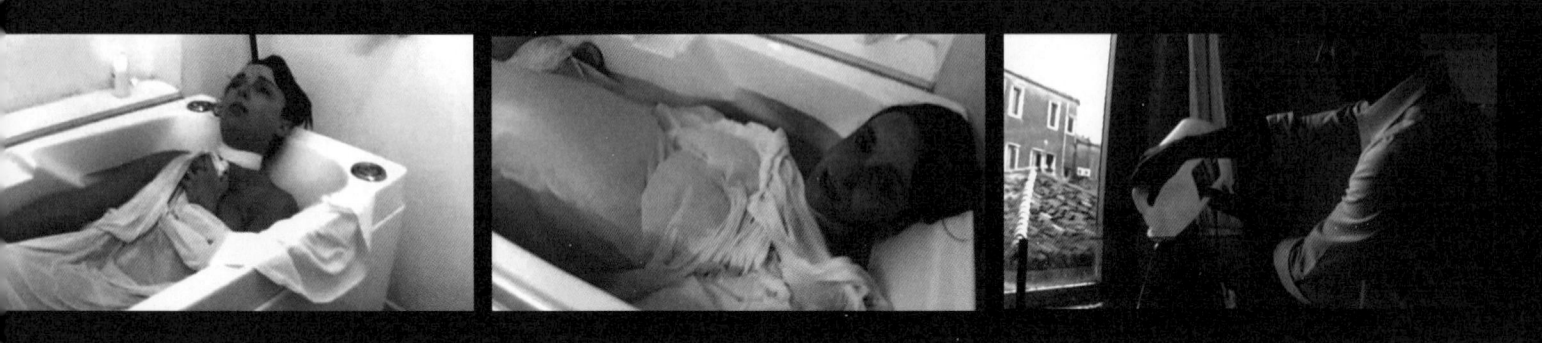

Screen shots from Lech Majewski's movie *The Garden of Earthly Delights*.

The Complexity of The Human And The Complexity of The Urban

In Lech Majewski's movie *The Garden of Earthly Delights* (Ogród rozkoszy ziemskich, 2004), a terminally ill art expert, Claudine and her lover, Chris travel to Venice. She needs to give a lecture on Hieronymus Bosch's famous painting there, but neither the immersion into that incredibly rich work of art nor beauty of the unforgettable city can provide more than a backdrop for yet another death in Venice. Majewski's plot and swinging, hand-held camera masterfully blend the depths of one of the most amazing urbanities, the vortices of Bosch's imagination and an imminent end of life, a life at its loving high. An air of irrevocable ending sets a melancholy undertone which tames all the excitements which an extraordinary place and the best that art and human feelings could offer.

During one of her emotional lows, caught in the bathroom, with sounds of the church bells in the background, Claudine contemplates materiality of her frail body. Chris goes to a grotesque extreme and sets up an elaborate mock-up of ingredients which make human body. An aquarium full of water: "There, this is exactly thirty eight liters of water, which makes up three quarters of your body." A pile of white crystals: "Over here, I've got three kilos of ammonium nitrate, that was the only way to get hold of nitrogen. Then, we've got ten kilos of carbon; this is what your body uses for energy. Then here, this ordinary blackboard chalk is your calcium content, exactly. This is what makes up your bones. This is three grams of iron. And this, this rivet, is exactly one gram of zinc; this is to help your digestive process. And then, finally, this paper clip is exactly one fifth of a gram of copper. This colours your skin." Gazing through the window as everydayness of Venice unfolds and moving her fingers over those materials, Claudine says to the engineer she loves: "So, this is my portrait", only to make her lover turn his hand-held camera back, towards her: "No, here we have the real you, Claudia, in the flesh …". She bends, kisses the camera lens, she coughs. The movie ends soon after. The limits and finality of human existence.

Claudia, in the flesh. The complexity of human being exceeds our capability to comprehend and represent its many realities, but that never stopped our efforts to seek and communicate its elusive essence. The complexity of a human being, of each and every human being is simply incomprehensible. While our bodies do consist of all those elements and materials, while we indeed *are* physical, we are at the same time so much more. At least, the Cartesian *thinking bodies*. Each of these bodies in the street, in the metro, comes with an incredible web of connotations, links and interconnections, entangled in various histories and memories, knowledges and feelings, faults and beauties, rhythms, rhymes, passions, secrets.

We are material, and – more. Always, much more.

Now, let's imagine an effort to represent Venice, in a way equivalent to that in which the engineer, Chris, started responding to Claudine's wish. I am absolutely sure that we are technically capable to calculate the exact volume of water in the Venetian Lagoon, at any moment in time. We must also be able to say exactly how much mud and who-knows-what-else makes the many islands of Venice, how many millions, billions of tons of various materials are embedded in all of the houses, bridges and squares which physically make that unique city … Possibly down to a kilo. (If not yet, then that is only a matter of time.) But, would that quantity be – Venice?! "So what?" is the most damning of questions confronting our immense power to quantify.

One of my favourite definitions of *the urban*, favourite as it is at the same beautifully useless and the most comprehensive definition of all, is Oswald Spengler's declaration that the city is *a settlement with soul*. Those words resonate long after the overall relevance of Spengler's overall opus has vanished. He, significantly, points out how "the real miracle is the birth of the soul of a town. A mass soul of a wholly new kind … Out of the rustic group of farms and cottages, each of which has its own history, arises a *totality*. And the whole lives, breathes, grows, and acquires a face and an inner form and history. … It goes without saying that what distinguishes a town from a village is not size, but the presence of a soul" (Spengler, 1928).

That *mass soul* is the *soul of culture*, which I associate and, on occasions, equate, with – *urbanity*. But here, we are interested in introduction of *totality*, that elusive quality of identity-defining intensity which explains the key difference between the "village" and the "town". Crucial difference is not in size, nor in any other easily identifiable and measurable aspect, but in the level of complexity. The *miracle* indeed is in the capacity of *the urban*, comparable to that of *the human*, to transgress the physical, and to reach an intangible, but all-important quality – identity. In Spenglerian terms – *the soul*.

Cities are factual, and – more. Always, much more.

The Method – Complex Approaches to Complex Phenomena

Lefebvre's recognition of an irreducible oeuvre resonates with the sense of totality of *the urban*, with that same, inexhaustible, *much, much more*. Lévi-Strauss argued for the need to uncompromisingly address complex, *total social act*, while best strategists of production of urban space, such as Alexander and Jacobs, also aim at the elusive, ideal, *whole of the urban*. For de Certeau, the city is the *most immoderate of human texts*, one that would pose the ultimate challenge for Hélène Cixous' way of reading, which focuses "not on a strategically selected detail but on *the text in its entirety*" (Andermatt Conley, 1992; my italics).

The latest developments in life sciences confirm the necessity to deal with complex situation in their entirety. Complex realities need intellectual apparatus of matching complexity as, in order "to begin to understand many aspects of our complex world ... we need to expand our conceptual frameworks to accommodate contingency, dynamic robustness, and deep uncertainty" (Mitchell, 2012). While quantifiable

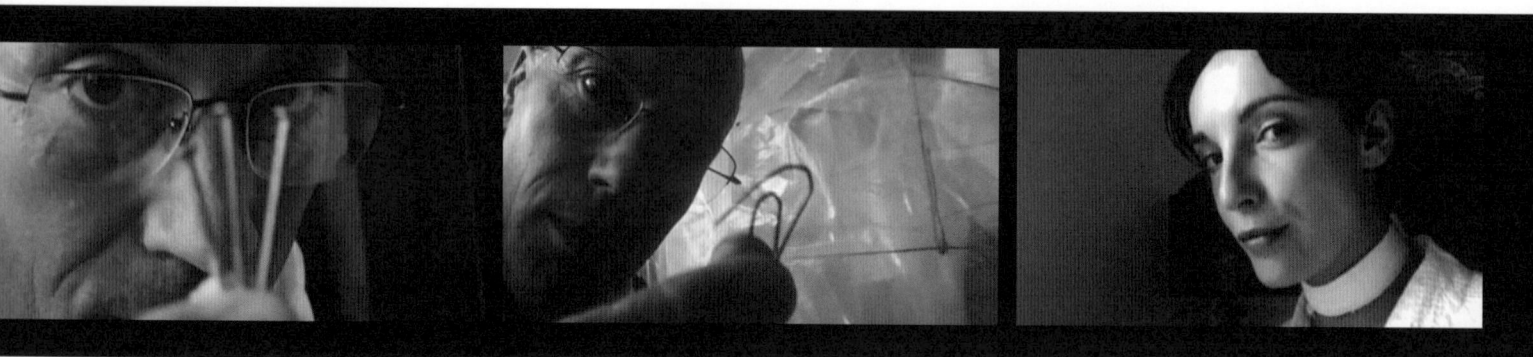

Screen shots from Lech Majewski's movie *The Garden of Earthly Delights*.

and for bad, came with Lavoisier and introduction of precision, quantitative measurements and the idea of modernity, not only as different, but as decidedly superior to previous epistemological frameworks. His enormous opus has "constructed a completely new language that rhetorically organized the world in keeping with those (new) views, and set out to convert others to their program ... Lavoisier rhetorically linked theory, experiment, and instrument into an indissoluble whole. Acceptance of his instruments implied acceptance of how he structured chemical investigations generally" (ibid.). Previous knowledges were totally excluded, the singular replaced the plural in order to enthrone a new, "pure" epistemological framework, an exclusive paradigm of modernity. The emphasis started to move from cumulative and evolutionary, towards futuristic and revolutionary. In the world of science, "where Rouelle engaged the senses in a search for qualitative distinctions, Lavoisier looked largely to measurement for experimental determinations that supported his theoretical claims" (ibid.).

Life sciences find such limitations too restrictive and unsustainable. The problem is even more pronounced in the studies of built environments, as they include the complexities of both ecological systems and those of human and social power relations, where exactly the latter produce crucial quality, the layer that makes that *human text* of *the urban* so amazingly *immoderate*. Approaching those realities "requires, in many cases, a more explicit and detailed analysis of the many roles context plays in shaping (natural) phenomena. It means that conditions often relegated to the status of 'accidents' or 'boundary conditions' be elevated to the subject of scientific study" (Mitchell, 2012). If we allow ourselves to lose sight of the complexity of urban phenomena, of the marginal and liminal, if we agree to reduce our understanding of, and our interaction with cities, if we narrow them to utilitarian sets of fragments and fragmentary solutions, we will inevitably lose ability to think, to make, and to live *the urban*. We will lose complexity, as one of the key urban features in itself. We will lose urbanity (which cannot be fragmented) and, ultimately, human dimension of our habitat.

dimensions of life matter enormously, they cannot cover the overall, synthetic totality. Contribution of the numerical remains invaluable when it comes to many aspects of the investigated phenomena, to a degree defined by concrete purpose, and within the constraints of unstable hierarchies of instrumentalised facts. As we have seen in the example of Majewski's lovers, facts alone are not sufficient when it comes to capturing the defining quality of a single human being, let alone that of the collective socio-cultural "being". Cities are always in, and of a particular place, in and of a particular time. Such double contextualisation makes their realities enormously dynamic and complex. The complexity itself and the groundedness in concrete, unique situations are the key aspects of *being urban*.

Until recently, much of sciences were unable to deal with complexity of their fields of investigation, as they, "especially prior to the latter part of the twentieth century, adopted strategies involving reductive explanations designed to simplify the many complexities of nature, in order to understand them" (ibid.). An interesting illustration of the emergence of reductionism in science may be found in transformation of pharmacy in the 18th century France. Rather than seeking general laws, traditional chemists such as Gabriel-François Rouelle, the "task was to analyse ... peculiarities and generate individual facts". Training the senses was crucial to the process and purpose of chemistry ... Once trained, the eyes, nose, ears, hands, and tongue of the chemist would guide him through the richly heterogeneous world of nature, delineating substances by their sensibly observable characteristics and effects" (Roberts, 2005). Such was the case with many traditional medical systems, including the rich lineages of Chinese *zhōng yī*. The advancement has, gradually, taken the personal and sensual out of science. Bringing the latest knowledge in evolved into an absolute exclusivity of the new way of looking at the world. Dramatic change, for good

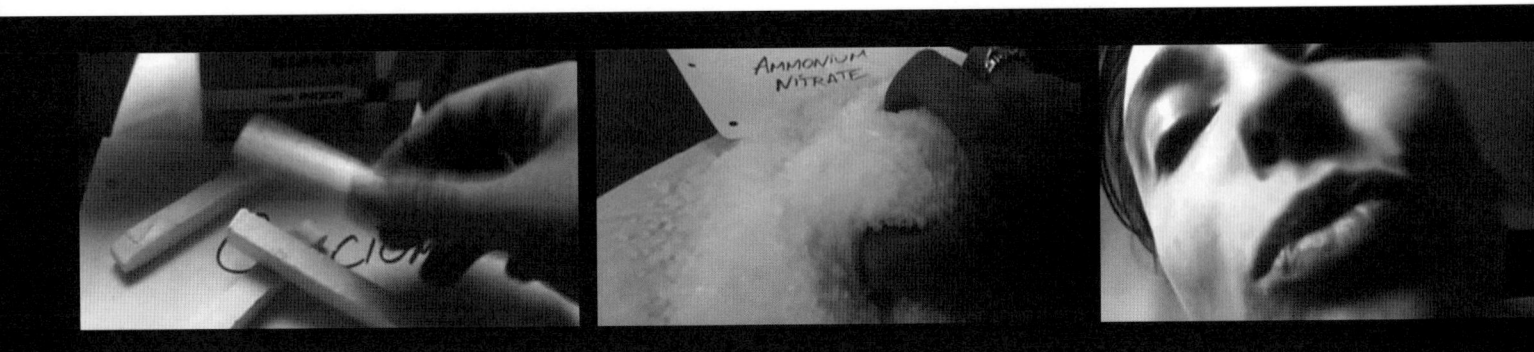

In that sense, the complexity of approaches needed to address *the urban* means precisely (re)discovery of *fully human* ways of thinking and making cities as good, profoundly *humane* urban environments.

Exclusion – And What Gets Taken Out from Urban Research

It is interesting to see which urban dimensions tend to be left out, or proclaimed irrelevant by standard research practice.

Favouring the factual and numerical over the elusive and textual, the guardians of the ruling paradigm carefully identify what can not fit the existing power structures (of knowledge). That includes various non-measurable, non-scientific, bottom-up, subjective and sensual knowledges – in short, the untamable and thus proscribed. We could call such ostracized practices *heretic*, remembering the etymological roots of that term, Greek *hairetikos*, which referred to those who were "able to choose". All power structures (which include those of knowledge) tend to restrict choices. That tends to apply to practices with potential to undermine the ruling value system. Umberto Eco usefully explained how the usage of the term *heretic* evolved, from meaning *those who can chose the way they think*, to *those who think differently*, only to end up meaning *those who think differently, and therefore – wrongly*.

In terms of the main focus in this book, rigorous application of rigid research frameworks excludes many subtleties which make some of the most fragile, most beautiful and most precious dimensions of *the urban*. Even mentioning beauty in the context of research, came under suspicion as, ultimately, subjective and thus impossible for pedantic processing and classification. The same applies for multisensory experiences, where various hierarchies gradually turned inflexible and reductive, ending with an almost total favouring of the eye.

The loss of experiential depth (Harvey, 1990), which came along with the modernist urge, is favoured and generously supported by the power of *Spectacle* (Debord, 1994; 1998; Wark, 2013). "The only sense which is fast enough to keep pace with the astounding increase of speed in the technological world is sight" (Pallasmaa, 2005). The only power which was capable to catch up and take over that condition was the unrestrained power of the "free" market Žižek, 2005). The *Spectacle* cherishes "the world of the eye", as it is "causing us to live increasingly in a perceptual present, flattened by speed and simultaneity" (ibid.), in which its artificial realities of money and globalized financing operate best. Domination of the optical introduced a particular kind of "hygiene" which fears complexity. "The contemporary city is increasingly the city of the eye" (ibid.). It superseded the "haptic city" of the past (ibid.), producing the shopping-centre-like sterilities, which get confused for *the new urban*(ism). Culture – and place-specific shadows of Tanizaki, Rembrandt or Caravaggio give way and open our realities to pornographic gaze, surveillance, and characterless, industrial lighting.

Since economy has been proclaimed a science, and especially after it became equated with politics, the logic of a single bottom line of the "free" market has conquered the world. Such economy-cum-politics, governed by the *Spectacle*, ensures power for the few, and provides *panem at circenses*, enough food and entertainment to its subjects. That power has constructed its own language that has, in that historically proven, Lavoisier's way, rhetorically (re)organized the world in keeping with its own (new, always new) views, and set out to convert others to their program. The unconverted ones are, of course, *heretics*.

As Majewski lets Chris explain, and as we all know all too well, the most important dimensions of our reality, its "events and processes are not simply complex in the sense that they are technically difficult to grasp; rather, they are also complex because they *necessarily exceed our capacity to know them*" (Law, 2004; my italics). When emphasising how the events and processes necessarily exceed our

Fragments from Darko Radović's lecture.

While the *Spectacle* leaves false impression that "anything goes" (as long as the markets are "free"), the harsh realities and inequalities it produces tell exactly the opposite. Power which fuels neo-liberal globalisation is very efficient in exterminating ideological opposition. The kind of urban inclusiveness which we advocate here is not some extension of the Post Modern "non-ideological", banal pastiche. To the contrary, it has to be based on strong values, which revolt against, and are in opposition to the ruling doctrine. This value system rejects its own ossification into a new totalitarising ideology. Getting there demands new kind of thinking and decisive individual responsibility.

The urban is inevitably ideological and political. Ideology and politics of *the urbane* are those of *common good*. Over the last several decades, *the urban* has been reduced to banal urbo-economics, in parallel to, and as yet another expression of the reduction of citizens to mere consumers. Therefore, what gets taken out from urban research and action is the fullness of our humanity and the awareness that, in order to reach our individual and social selves we need (to demand) *humane* environments which can nourish the best we can all give and live.

In what follows we will focus at two dimensions of urban research which have been neglected and ostracized, as they could not fit the insatiable growth machine of the *Spectacle* – our subjectivities and our sensualities. Subjectivity and sensuality are political and subversive. The *Spectacle* banalises and reduces sensuality to sexuality, and sexuality to pornography; subjectivity to individualism, and individualism to selfishness. In that way, it reduces everything to the (largely monetary) numerical. The opposing project is in search for the complexity lost.

Subjectivity –
And How to Include It in Urban Research And Practice

In order to conquer everyday life, and diminish individual human capacity to know them, Law actually stresses the beauty of being in the world, the very existence of the unknowable – the search for which, besides cogito, makes us human. Sentio ergo sum. I feel, therefore I am. Facing the limits of our ability to rationalise demands courage, courage of an even higher order than what is necessary to unveil the truths (or – the "truths", scientific – or otherwise).

An ability to appreciate, in parallel and in addition to the need to comprehend (and not as an replacement for it), is exactly what has been lost by application of exclusive, progressist methods in addressing the aporias of the urban. That comes from an overall unease of researchers to face situations that can not be tamed and conquered to fit dominant epistemological frameworks. They "direct themselves at that level of the totality over which their discipline claims proprietary rights. Thus the disintegrating spectacle finds itself confronted with fragments of specialized knowledge that cannot but think on the fetishizing terms

of the disciplines that birth it" (Wark, 2013). Despite the rhetoric, the power which fuels current globalisation has Lavoisierian difficulties to acknowledge even the existence, let alone the importance of *tout autre*, a true and radical difference (Derrida, 2006). That is because such otherness "cannot be made transparent to the understanding and thereby dominated and controlled" (Hillis Miller, 2001). As any heresy, otherness is seen as subversive and, consequently, labeled highly undesirable. On the other hand, *the urban*, as an essential theatre of co-presences, is exactly where the *otherness* of *the Other* plays a significant role in creating that magic, irrepressible complexity that makes cities. That includes some of the most subversive (urban) rights – such as Lefebvrian *droit à la ville*, the renewed right to the city, *le droit à la différence*, the right to difference ie. the right to be empowered and to be different, and *rights to (each particular) city and to (each particular) urbanity*, as fully-developed local cultures (Radović, 2008).

potentials and collective urge to meagre instrumentalised consumerism. The *Spectacle* reduces experiential depth of our existence, by favouring flattened images over profound realities. It favours nuclear individualism, which detrimentally affects *the urban*, as the most profound spatial projection of social essence of human being.

Our existences, as we are going to see, are profoundly relational. Everything about us is *au pluriel*, and thus about the relationships between various, and variously interacting subjectivities. In urban research and production of space, that means simultaneous singularities of all actors involved. In architectural design, subjectivity is accepted and even supported only when it is about artistic, and thus highly exclusive dimensions of space. That, to a certain degree, extends to urban design, confining it to the same mould of exclusivity. Further from the exclusive realms of design, towards what should be inclusive processes of (urban) planning and research, subjectivity – as it can not be easily controlled, gets strictly proscribed.

The urban can never be subjected to any single knowledge or power, despite all the efforts to do so. There were never true "owners" nor true "authors" of any proper city. "A private city" would be an oxymoron. Cities are always layered realities which, at all lavels, include multiple, entangled subjectivities. The synthetic flux of those realities, the actual urban fabric, its real places and times, is always as much accidental as it is the product of careful planning and development strategy. That is due to the immense complexity it embodies. The reason why the Spectacle so desperately seeks to flatten and simplify urban realities is in order to control *the urban*. That is also one of the reasons of perpetual urban crisis. All efforts towards flattening stay pitifully unsuccessful.

As Guy Debord came to recognise, "the essentially personal nature of the relationship between the individual and the city, sensing that this subjective realm was always going to remain at odds with the objective mechanisms of the psychogeographical methodology that sought to expose it, 'The secrets of the city are, at a certain level, decipherable,' wrote Debord, 'But the personal meaning they have for us is incommunicable.' " (Coverley, 2006). Dialectics between the *decipherable* and the *incommunicable* in the urban is precisely where the key argument of this book is placed.

Such epistemological frameworks are never "pure", as there are no "pure" realities in *the urban*. Correspondingly complex methods also need to be "impure". From such impurities often emerge those missing, *incommunicable* qualities. In their "impurity", these frameworks tend to have complex and composite foundations, as in Lefebvre's or Jean-Luc Nancy's intellectual lineages. In the case of the latter, for instance, inspirations are marked by thinkers as different and (in this sense) as (in)congruent as Rousseau, Marx, Freud, Heidegger, Husserl and Lacan. In urbanism, in a comparatively complex way, there is no overarching, successful theory but always – theories, plural. Those theories tend to be fueled by the power of *singular* thinking, which lets itself be layered over and below all other attempts at thinking and making *the urban*. As Derrida reminds, "scientific theories are described as *commensurable* if one can compare them to determine which is more accurate; if theories are *incommensurable*, there is no way in which one can compare them to each other in order to determine which is more accurate" (Derrida, 2005; my italics). Accuracy is often not at stake here; preferred might be blurs or, as we are going to see later – the knots.

In the context of this approach to *the urban*, as for Jean-Luc Nancy, singularity "is not individuality; it is, each time, the punctuality of 'with' that establishes a certain origin of meaning and connects to an infinity of other possible origins ... The togetherness of singulars is singularity 'itself'" (Nancy, 2000), which means, in a way, that the very *complexity of inter-subjective is the smallest unit of culture*! We need to remind the reader here that Greek term for an individual – defined as a private

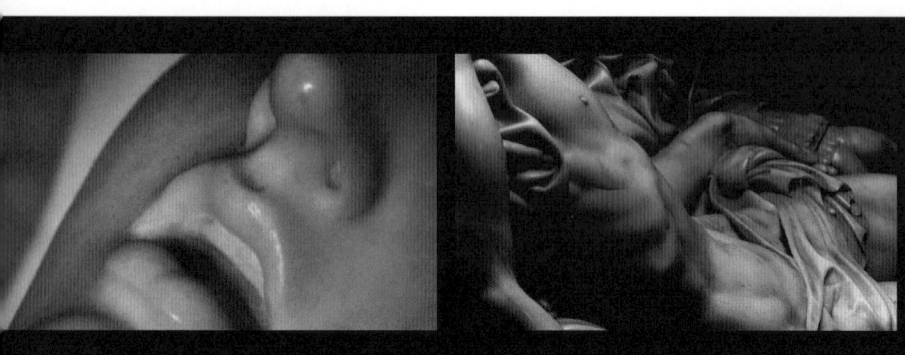

Fragments from Darko Radović's lecture.

in a total way with *my* whole being, which speaks to *my* all senses at once" (Pallasmaa, 2005, my italics). That brings us back to Spengler, Lefebvre, de Certeau and others from the opening pages of this text. When they speak about *totality* of *the urban*, they speak about an ultimate complexity. That is not the complexity of extraordinary places and moments, but that of everyday life, of Barthesian rien (Barthes, 1982), the banal "nothing" which, albeit unpresentable and incommunicable, in Bachelard's terms, makes the "polyphony of the senses" (Bachelard, 1996). Much of our *being with* is immersed into, and defined by that polyphony of our making.

Ordinary, banal, quotidian qualities (perhaps, surprisingly for some) *are* the ultimate complexity of cities. Everyday life and the *Spectacle*, are irreconcilable opposites, one being the world of rich singular plural identities, and the other the world of self-sufficient individualities (or, if one prefers Greek, of *idiotes*).

Urban design of the kind we advocate here is not only the practice of thinking and acting (with)in *the urban*, but also a particular sensibility and an ability to reach into the human(e) core of cities. We are speaking about multiplied, layered subjectivities in research and design in which, rather than flattened, we want to keep each and every singularity of the participants identifiable and relational. The resulting palimpsest may not be simple. It indeed is *immoderate* (de Certeau), *unpresentable* (Derrida), *incommunicable* (Debord), non-*measurable* (*Mn'M*), as each scratch, every line or trace of the brush on it matters, or – it could matter.

If there is something to bring in from René Descartes here, that would be his emphasis on the importance of doubt. Nancy, addressing the issue of (self)confidence, speaks about "us, we who are supposed to say *we* as if we know what we are saying and *who* we are talking about" (ibid). That directly opposes the flatness and simplicity of *Spectacle*, the epitome of which could be found in any of the overtly self-conscious, but shallow exponents of venture-financing, that flagship of neo-liberal, spectacular culture. Doubt always was, and it remains at the source of knowledge. Here – not only of our

person, one not taking part in public affairs, was – *idiotes* (which later, via Rome and Mediaeval appropriations acquired its contemporary meanings and which, today, in a quite convincing way, could be linked with the idea of a perfect consumer of the *Spectacle*.

That is so because in a *human* and *humane* world, "being" is communication (ibid.), as our being is *un pluriel d'existence*. The worst punishment a man can inflict of the other is solitary confinement, a worst nightmare – an deserted island.

Our existential urge is to create strings of meanings and bring some order into realities that we live. That is how we create places, building them as meanings, which take various physical forms (or the other way round), as in Heidegger, on whose shoulders Nancy wants to stand. Or, in shortest (not as a summary of Jean-Luc Nancy, as his thought could never be summarised. Nancy, at best, can only be accurately quoted), "*the plurality of being is at the foundation of Being.*" (ibid.; original italics). Beyond that "unit" of being, " 'with' is the sharing of time-space; it is the at-the-same-time-in-the-same-place as itself, in itself, shattered. It is the instance scaling back to the principle of identity ..." (ibid.). Cities are the embodiments of human need for coexistence, with others of our own ilk and with the radical *Other* (including that within our own selves).

As in Debord, whom Nancy has read and interpreted carefully, urban quality evades us as "the *co-* of copresence is the unpresentable par excellence, but it is nothing other then ... presentation, the existence which co-appears." (ibid.). The *unpresentable*, that ultimate quality, matches Debordean *incommunicable*.

A devout anti-Cartesian, Maurice Merleau-Ponty wrote: "*My perception is* [therefore] *not a sum of visual, tactile and audible givens: I perceive*

stone | subjective | deeply personal

cerebral knowledge, but also of our variously suppressed, subliminal knowledges, which all would, ideally, add up to an impossible and ideal, *total* knowledge.

Those diverse knowledges are exactly what have been spaced out from urban research and action, since the neo-liberal globalisation took over. What we have lost is an awareness that "all of the being is in touch with all of being ..." (Nancy, 2000) or, shorter, we have lost our own humanity.

Sensuality –
And How to Include It in Urban Research And Practice

What has been spaced out was precisely, as in Lavoisier laboratory, all that was judged impure, potentially dangerous, sensual, all that is lurking under the surface of *Spectacle*.

Hierarchies of senses have been sought at least since Ancient Greece. During the Renaissance, the privileging of the eye evolved into a fully developed "ocularcentric paradigm and strong, dominant, anthropocentric worldview has emerged. In terms of production of space, ocular-centrism reached its dystopian pinnacle in the 19th century, with Panopticon, a perfect prison with the building itself an ultimate control tool. Since Jeremy Bentham's times, we have seen only exponential rise and perpetual fostering of that logic.

But, despite the aggressive globalisation of the logic of Panopticon, the mainstream has not managed to suppress the heresy. Not yet. The rebellion starts with an irrepressible sensuality, with hand – "or, perhaps, more exactly with the *finger of the hand*" (Derrida, 2005). Subjectivity can never be dislocated from the body, from each and every concrete, sensing body. That is especially the case when subjectivity is understood in Nancian terms, as singular plural, as being-with – which is critical for comprehension of the essence of *the urban*. There is an urgent need to reconquer that complexity in order to deal with many manifestations of urban life, such as alienation (one of many terms that are, as uneasy and difficult to address, not popular in the dictionary of the happy *Spectacle*).

Being in the business of place-making, urbanists and architects need to know that "bodies are places of existence ... (that) the body *makes room* for existence ..." (Nancy, 2008). "The body of sense" is polysemous (as Nancy's term itself is full of meanings, ranging from an understanding of the body as locus of sensation, seat of sensibility, to our [impoverished] common sense). "The body is being-exposed of the being" (ibid.). Its existence is spatial. "Bodies don't 'know', nor they are in 'ignorance'. They're elsewhere, they're from elsewhere, from another side (of places, regions, frontiers, limits, but also of household plots, boulevards with promenades, trips through estranging lands: in fact, they can *come* from anywhere, from the spot, even *here*, but never from the nonplace of knowledge)" (ibid.). Immersed in that, being one *with* that, the questions arise: "Why is there this thing, sight, rather than sight blended with hearing? And would it make any sense to discuss such a blend? In what sense? Why this *sight*, which doesn't see infrared? This hearing, which doesn't hear ultrasound? Why should every sense have a threshold, and why are senses walled of from each other? Further still, aren't senses separate universes?"(Howes, 2005).

Addressing such questions, Michel Serres (2008) speaks about knots of our sensual experiences, complex entanglements which we have walled off in our desire towards clarity, commensurability, deciphering. Coverley finds something similar in De Quincey, who wrote how in his urban wonderings he "came suddenly upon such knotty problems of alleys, such enigmatical entries, and such sphinx's riddles of streets without thoroughfares, as must, I conceive, baffle the audacity of porters, and confound the intellects of hackney-coachmen. I could almost have believed, at times, that I must be the first discoverer of some of these

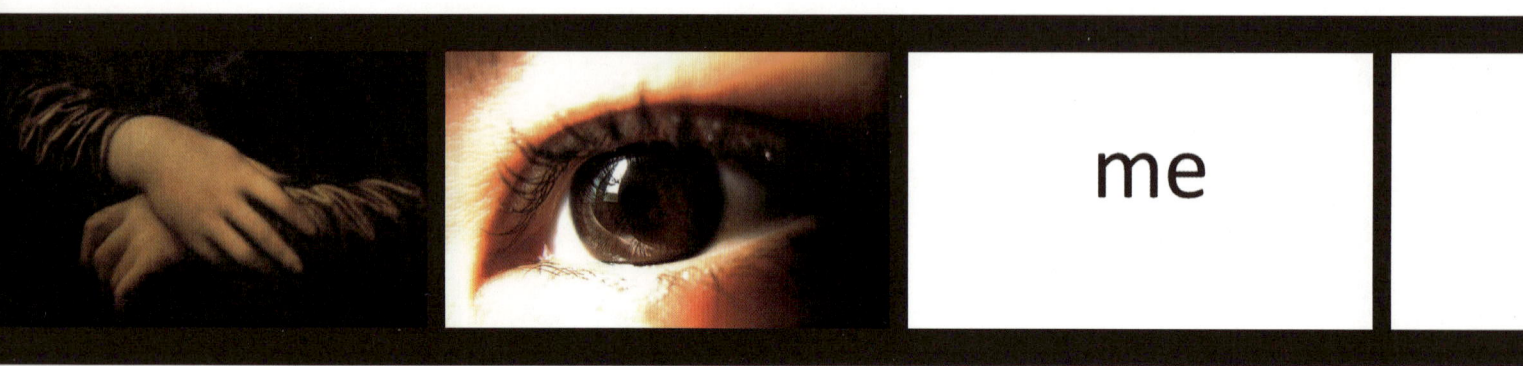

Fragments from Darko Radović's lecture.

For Merleau-Ponty, "every sensation is spatial" (Derrida, 2005). I add: every spatiality is, equally, sensual. Spaces are knots of sensual stimuli, sensations are knots or perceived spatial realities, in which "the senses communicate with each other" (Merleau-Ponty, 2012). They reduce us to creating the self-gratifying, simple and easily understandable constructs – which, alas, usually do not have much to do with realities they are meant to represent.

In summary – explorations of space need to include sensual, which is, by definition, subjective. Presenting those dimensions and opening them to thinking, including analytical thought, needs to move from (de)layering and (re)layering of realities, and include (un)tying and (re)tying (and tightening) of the knots of sensations of *the urban*. That, of course, is an utterly impossible task, which makes it worthy of investigation, and its application promising!

Dérive, or –
How to Seek, Find and Capture Subtle Urban Qualities

In order to pay justice to our own sensual selves, and in that way to our own, *full* humanity, we need "new" dimensions of knowledge. Inverted commas on the "new" here are deliberate. Those "new" ways often include the return of ostracized subjectivities and heretic sensuality, along with recognition that, in parallel and in addition to our standard(ised) analytical faculties and methods, we need Bergsonian "intuitive vision (which) does not just come into contact, as it is said; it *becomes contact*, and this movement would pertain to its nature" (Derrida, 2005). Artists would not need an explanation of what is meant by that, but exponents of the compartmentalized knowledge probably do. What is needed is akin to those practices of traditional sciences as in Rouelle's laboratory, which favoured a person to person contact and "the more in direct touch they were with the world of their investigation, the less inhibited they seem in comparison to current standards of hygiene and gentility" (ibid.).

terrae incognitae, and doubted, whether they had yet been laid down in the modern charts of London" (Coverley, 2006).

That makes perfect sense. While seeking complex ways of thinking and making cities, we have discovered the ultimate complexity of the object of our research. Appropriate, complex knowledges can not be made simple, unless we decide to accept the simplistic thought. Layering of the textual – simplifies; textuality of the messy palimpsest does a bit better; and knots elevate the discourse yet another notch up. All of those combined reach towards *vécu*, fully human, fully lived realities of everyday life. Although not necessarily open to ratio, representing those realities helps open them to alternative ways of comprehension, which can be usefully translated into inspiration. As Hélène Cixous wrote in *Savoir*, "what is given to see is given to touch, though henceforth, from the outset, it is given to read" (Derrida, 2005).

But, what is given to touch is given to an *open* reading, to various sensual and subjective appropriations. That is what systems of control do not like, and that is why they prefer standardization and accuracy, if possible, a scrutiny of the panoptically positioned eye. "Lower senses" get appropriately excluded, as "non-measurable" rendered inferior, at best only referential. Our proposal here is to seek fully multisensorial practices of urban research and action. They demand multiple entry points, various planes and plateaus (a la Deleuze and Guattari's, 1987), rhizomes which are "against method" (Law, 2004), and thus – resielient. The key questions asked indeed move us from "What does it mean?" to *an inclusion* of "How does it mean?" (I never understood why in Deleuze there is an excluding move *from* "What does it mean?" *to* "How does it mean?" (West-Pavlov, 2009), and not an inclusive *both-and*.) That is where we (can) understand (urban) qualities as (diverse) *intensities*.

Urban research which celebrates rather than oppresses subjectivity and sensuality of its subjects and exponents needs genuine, person to person and person to place contact, direct touch.

A decisive move in that direction demands (1) reevaluation of the role of senses (a missing step towards recognition and comprehension, and attempts towards representation of the *totality*), (2) reintroduction of subjectivities (as an opening to participation, inclusion of lived experience, an all-important vécu from the Lefebvrian triad of conceived, perceived and lived space, Lefebvre, 1996), and (3) complexity of action (as intrinsic to the discipline of urban design, which balances the strict exclusivity of *design*, with generous inclusivity of *the urban*; Jacobs, 1989). While those requirements which might not be new, their very recognition would amount to a revolutionary paradigm shift. The necessary methods are both extensive and intensive, rational and intuitive, cerebral and sensual. They demand an active immersion into the phenomena explored, eye-level, hand-level, nose-level sensing,

touch. Involvement in the everyday, *vécu*. Their implementation needs to include standard data, as well as layers and knots of variously subjective sensations. Such methods seek smooth flows, without avoiding eddies and not glossing over the striations. They need sensibilities well captured in the Benjamin's *Berlin Chronicle*, where he wrote how "not to find one's way in a city may well be uninteresting and banal. It requires ignorance – nothing more. But *to lose oneself in a city* – as one loses oneself in a forest – that calls for quite a different schooling" (Coverley, 2006; my italics).

In *Mn'M* project, I explored the potential of renewed, reframed Situationist *dérive* (Debord, 1959) as "a mode of experimental behaviour linked to the conditions of' urban society: a technique of transient passage through varied ambiances" (ibid.), aiming to combine Debord's ability to be simultaneously free and rigorous, with Perec's sensibility and an eye for detail. As Bellos explains, Debord and Perec are far from incompatible. "Perec may have heard of the (Situationist) group through Henri Lefebvre, who had

had the good and the bad fortune to be Debord's teacher in the 1950s; or he may have come across one or another of its irregular publications in *La Joie de Lire*; or he may simply have picked the Situationist notions that were in the air" (Bellos, 1993). Perec shared our liking of Situ's argument "for a renewal of the ways in which the world was seen", as they put forward "two 'techniques' for setting things aright: *la dérive* and *le détournement*. A *dérive* (literally a 'drift' or 'drifting') was an aimless, rapid perambulation intended to reconfigure the urban environment and allow the shape of things to determine situations: instead of using the city, the "drifter" allowed the city to use him or her. *Détournement* ("kidnapping", "subversion", "misuse") was the active corollary of dérive" (ibid.). Theirs was Deleuzian smooth space, "explored without calculation, without being qualified, ... explored as a body of 'rhizomatiques' which are explored in the moment of travelling ... space of nomadism is tactile, haptic, sonorous ..." (West-Pavlov, 2009). That is what we seek when searching the ways to empower our subjective and sensual selves, expose alternative knowledges to *the urban*, with an ultimate aim of freedom, freedom *to be urban*, freedom for *urban being* (as, indeed, *Stadtluft Macht Frei*).

In *dérive*, we speak about an *immersed* body, an immersed self (always a concrete self, somebody's *my*-self), an experiential and thus subjective (ad)venture – adventure of *my* own mind, *my* own senses, thus sensual – of *my* own body, *my* own mind. There are some fine resonances of that in Husserl's "rigorous phenomenological method", which celebrates "immediate accomplished, and 'perfect' intuition of the physical; 'originary presenting intuition' (in our case, experience), the sole founding ground of a theory even if this theory is determined as predicative in mediate thinking" (Derrida, 2005). An ability to operate in that register needs a new way of thinking, one which liberates, rather

Fragments from Darko Radović's lecture.

The key to successful *dérive* is freedom. It is very important *who* takes the courage to drift and to "let go". In the case of seasoned urban researchers, embedded knowledge provides solid foundations that enables dialectisation of spontaneity, diverse subjective fantasies, and groundedness in relevant ideas and established knowledge. Thus, "the randomness of a *dérive* is fundamentally different from that of the stroll, but also ... the first psychogeographical attractions discovered by *dérivers* may tend to fixate them around new habitual axes, to which they will constantly be drawn back" (Debord, 1959). As learned during the Mn'M Tokyo *dérive* experiments, it is of crucial importance to take off the pressures of the *habitus*, in this case the straitjacket of knowledge cut by the ruling paradigm. That includes the attainment of freedom from standard research practices, abolition of the baggage of officially sanctioned theory and/or practice, lose focus and free selection of concrete spaces and practices for physical and intellectual immersion – as there is the possibility of liberation, of that frightening freedom (Radović, 2013a).

Essayistic Sensibility, or – How to Communicate the Findings

As suggested before, representation of subjective and sensual knowledges, when possible at all, does not necessarily demand precision and accuracy. Even when they allow elaboration, those knowledges do not inevitably lend themselves to explanation. But, as only true knowledge is shared knowledge, sensual findings need to communicate (the *incommunicable*). They have to trigger, to inspire, to help intuit further knowledges and action. Those are always singular knowledges, emerging from subjectivity of *my* own mind, sensuality of *my* own body. Representations of such knowledges need an essayistic sensibility, they have to keep doubt, enter aporias which highlight the *non-finito* nature of a lived intellectual (ad)venture.

Theodor Adorno's explanation of essay mentions speculative investigation of specific, culturally determined objects; intellectual and ludic freedom; childlike freedom that catches fire; aesthetic

than oppresses our other *faculties*. As "Heidegger regularly says (that) it is not because we have ears that we hear but, inversely, we have ears because we hear" (ibid.).

A famous palindrome *In girum imus nocte et consumimur ignil*, used as a title by Guy Debord for one of his movies – read left to right, or right to left, says the same: *we go wandering at night and are consumed by fire* (Radović, 2013). That encapsulates the essence of *dérive*, both as data an alternative gathering practice and as an interpreted, subjectified reality: an engagement with places *as seen, as experienced, as felt* by the one who is consumed by (her/his own) fire. The subjectivity involved, that fiery passion, confronts the very foundations of standard research practices, and exactly that sensibility may hold clues for (better) comprehension of the complex phenomena which tend to evade our scrutiny. Not accidentally, in his discussion of essay Adorno also refers to the "childlike freedom that *catches fire* (my italics), without scruple, on what others have already done" (Sheringham, 2006).

Debord's *Theory of the Dérive*, with precision of the manual (Radović, 2012, 2013a), explains the tactics of immersion into *the urban*: "*Dérives* involve playful-constructive behavior and awareness of psychogeographical effects, and are thus quite different from the classic notions of journey or stroll. ... In a *dérive* one or more persons during a certain period drop their relations, their work and leisure activities, and all their other usual motives for movement and action, and let themselves be drawn by the attractions of the terrain and the encounters they find there. ... the *dérive* includes both this letting-go and its necessary contradiction: the domination of psychogeographical variations by the knowledge and calculation of their possibilities." (Debord, 1959)

autonomy; own conceptual character; spontaneity of subjective fantasy; resistance to departmental specialization; realisation that there are 'little acts of knowledge that cannot be caught in the net of science; etc. (after Sheringham, 2006). Importantly for us here, Adorno insists how essays refuse 'airtight' concepts and embrace the changing and the ephemeral. He argues that, while operating in essayistic format, "the thinker does not think, but rather transforms himself into an arena of intellectual experience, without simplifying it' (ibid.)". For Georg Lukács, similarly, "the essay is a court, but (unlike in the legal system) it is not the verdict that is important, that sets the standards and sets the precedents, but the process of examining and judging" (ibid.). He emphasises simultaneous seriousness and lightness of essay. The culture in which knowledge is not the verdict, but the process, is the culture of tremendous intellectual generosity, one which allows ample space for the much-needed, and already emphasised, doubt.

Adorno also usefully reminds that "the essay's 'desire and pursuit of the whole' echoes the way all thinking about the *quotidien* involves a sense of whole – not an abstract totality, but a lived manifold of interconnections" (bid.) – as in Henri Lefebvre, Michel de Certeau, Oswald Spengler and many others. Perec, importantly, adds that essay rejects "the hostility to happiness of official thought' which seals it off against anything new as well as against curiosity, the pleasure principle of thought.' ... In its constantly self-reflective and self-relativising progress the essay co-ordinates rather than deduces ..." (Highmore, 2005). Proceeding 'methodically unmethodically', further provokes Perec, the essay becomes true in its progress, which drives it beyond itself, and not in a hoarding obsession with fundamentals" (ibid., and Radović, 2012).

As for Perec, for urban sensibility I advocate here: "*Space is a doubt*" (Highmore, 2005).

Materiality, or –
How to Represent the non-Representable

Since the times immemorial, since Imhotep, successful architects create beautiful, elegant, sublime spaces. The profession of architecture wrestles with the issue of (re)presentation of such, highly intangible, but critically imposrtant, qualities for millennia. The *venustas* in famous Vitruvian, tripartite definition of architecture – *utilitas* (variously translated as function, commodity), *firmitas* (solidity, materiality), and *venustas* (considered as beauty, delight, desire) – always had to be communicated in some way, and persuade those with means to translate an idea into actual, spatial, existential reality. Good architects produce spaces which we experience as beautiful, elegant, sublime, possessing exactly the qualities which are notoriously difficult to define (let alone measure) – but which we all, unmistakably, recognise. The question of what is beautiful is a minefield of subjectivities and Romans have, long ago, avoided argument by simply summarising how de *gustibus non est disputandum*, ie. declaring aesthetic taste beyond the bounds of rational debate. The same, of course, applies to senses of taste, smell, touch …, for much of what defines our environments and our being in the world. Putting any borders or definitions there is destined to transgress into highly subjective.

True architecture always was, and it remains schizophrenically torn-apart between its constitueing *techne* and *poiesis*. One of those two sides offers itself easily to the scrutiny of ratio, while the other strives on our feelings, intuition, sensuality and (suppressed) memories, reaching all the way to the exact opposite of rational analysis. And, the level of complexity of *the urban* is, again, of a much higher order than that of even the most complex piece of architecture. It reaches beyond an aggregate of architecture, into

Fragments from Darko Radović's lecture.

– nothing (modern times have forced film to lose even its celluloid body.) Even an expertly handled camera still captures only images and sounds. The lover remains unable to communicate the wholeness of his loved one – despite (possibly) being the only one capable of feeling the *totality* of the loved one's sensual self, the experiences which reach far beyond *cogito*, to *sentio* and beyond again, to the ultimate – *amo ergo sum*. Nancy has found, somewhere, how "Diderot claimed to envy painters, who could approximate colors, something he couldn't approximate in writing: a woman's pleasure" (Nancy, 2008). The elusive touch, touch of bodies, as the ultimate aim of representation of the sensual – for "the body is being-exposed of the being" (ibid.).

While their *dérive* included, and was significantly defined by sensuality of the participants, Situationist representations of their encounters with *the urban* also remain decidedly visual (often fascinating, but still only visual). With their extraordinary talents, some Situs, such as Asger Jorn and Guy Debord himself, managed to reach those repositories of our feelings, the fire within. In the fields of architecture and urbanism, mapping has established itself as a useful way to communicate, open the recordings to immersion and analysis, and thus extend our being-with. But, pushing towards an inclusion of subjectivity and sensuality, we need to ask – what is map? Does it accurately reproduce the territory? Can it capture the earthiness of "terra"? Does it give us 'the lie of the land', "a necessarily false and partial version of a place from upon which it claims offered a distanced, objective perspective? (West-Pavlov, 2009). Precisely to avoid such dilemmas, Deleuze and Guattari decided to use term *chart*, and distinguish between the implied utilitarian character for the *map*, and "imaginary, fanciful, indeed sometimes poetic qualifies may shake our belief in the representational veracity of maps. A *chart*, in Deleuze and Guattari's usage of the term, is open-ended, reversible, can be constantly modified" (ibid.). Those are the properties which the recording of the phenomena we are interested in has to meet. The *maps* which record subjective and sensual dimensions of *dérive* should be understood as "maps" only in the broadest sense, not necessarily referring to traditional, two-dimensional geographic representation at all. Those

an all-inclusive complexity, precisely (and uselessly, except for thinking) as defined by Spengler and de Certeau.

So, how to capture, communicate and (re)present these qualities? As Hélène Cixous suggests, "writing continues experience" (Andermatt Conley, 1992). (Re)presentations are such writings, and our writings of the city need to strive towards becoming direct extension of the phenomena explored. In *The Garden of Earthly Delights*, desperate to dispel any possibility that Claudine could be reduced to that pile of banal, measurable matter, Chris points his hand-held camera at her, letting her deep eyes and the sound of her name add the rest. Majewski masterfully blends sounds, mixes colours, textures and rhythms that hint at the tactility of place. He even suggestively hints at the smells of Venice, of love and of death but, in the end, both the lover and the master director had to scale the reality down to an animated image, a very complex one, but still only an animated image.

Media always subjugate our expression of realities to fit their own limitations (unless we accept that the message, indeed, is the medium). Great artists manage to overcome technical limitations by reaching into the repositorium of humanity of the viewers, or listeners of their creations. Jean-Luc Nancy explains that we scrutinize faces of the newly born children "with such curiosity, looking to see whom the child looks like and to see if death looks like itself. What we are looking for there, like in the photographs, is not an image; it is an *access*" (ibid.; my italics). That is the same urge which makes us travel, visit places, and then desperately seek how to sort out those new realities which we, somehow, already "know". A good book, a good painting or a fine piece of music, all have the capacity to connect with our innermost and let it compensate for what the shortcomings of the medium had to leave out. Otherwise, reduced only to the materials, a book indeed is just paper with some ink and glue; a painting, cynically, only a piece of canvas, with some chemicals applied to its surface; music or a digital movie

are, primarily, attempts to reach an essayistic charge in recording the places, the moments, the situations, the intensities which (can) trigger the feelings commensurate to those which have caused the act of (re)presentation in the first place. They, indeed, are akin to the fascinating nautical charts of indigenous Polynesian navigators, real *dériveurs* which, in a beautiful and *sufficiently* accurate way, tell the complex realities of the ocean and the sky, of waves and storms, currents and islands. One should remember that, for Debord, "from the *dérive* point of view cities have a psychogeographical relief, with constant currents, fixed points and vortexes which strongly discourage entry into or exit from certain zones. It is the role of the psychogeographer both to attune himself to these currents and to overcome them if they are to be reshaped and redirected" (Coverley, 2006).

These much-needed maps are acts of opening of the lived, or the experienced to knowledge. They need to be produced with an acute understanding that, whatever the medium, it can never capture full complexity of *the* urban. The processes of mapping and of reading such maps, as we have seen, extend interactions with the site, with places and activities of *the urban* – by other means. The hope is that, as such, they may attain the power to bring the reader close to the place and the moment where the researcher was, un-packing and un-flattening her his uttered messages.

The Scream, the Shadow and the Mirror

叫びと影と鏡像

Darko Radović | ダルコ・ラドヴィッチ

The Scream – A Note on Representation

Edvard Munch's *Scream* is, probably, one of the best-known works of art. Reflecting on its pastel version, the curator of the recent exhibition of Munch's work emphasised the startling power of that image, adding how, "almost despite the image's present-day ubiquity, ... the visual subtlety and complexity of this composition can't be summed up in a cliché" (MOMA, 2013). Celebrated for the combined power of colour and expression which communicate an overwhelming sense of unease and anxiety, the depiction of a human figure screaming in horror had many interpretation – with some reaching deeply into the artist's troubled private life, and others finding an equally possible collective unease of the late XIX century in front of the advancing Modernism.

But, our question here is: how did the painter manage to utter the scream on that mute, two-dimensional piece of drawing board? – as this is Edward Munch's own scream, regardless if those who see this work as a self-portrait are correct, or not? The power of the *Scream* comes precisely from deprivation imposed by the medium, from the impossibility that the sheet of paper could capture sound. Paradoxically, the very absence of its voice amplifies the cry, the howl of the hairless, asexual figure. Only implied, it becomes deafening. That voice, simply, *is* there, shattering our eardrums – by it's impossible absence! That face, the sheer horror of its expression, combined with the fire of the haunting sky and a strangely detached, uninhibited pace of the other two figures in the frame, tell us that that scream is there, so painful that it reaches far beyond what we are even able to hear.

An equally famous silent scream from Sergei Eisenstein's *Battleship Potemkin* (1925), also draws its immense power exactly from the absence of its main component – that of human voice.

The Shadow – A Note on Subjectivity

Shadows are never innocent.
Those painted by Giorgio De Chirico tend to be sharp, cold and dark. They reduce the visibility of shaded details. On the other hand, the shadows at the Impressionists' canvases are saturated with colour. They recreate reality and, in the process, they alter the "reality" they cover. Both kinds of painted shadows are said to be accurate; the sharp ones – geometrically true, the fuzzy-ones – closer to the reality of seeing. But, they both lie.
 Our eyes are never innocent.
The simplest of definitions of research sees that endeavour as production of knowledge, Etymologically, re-searcher seeks out, searches closely. What we see is, thus, strongly influenced by what we already know. Our knowledge is always there to conquer its subject.
 Knowledge is never innocent.
It carries the heavy *ideo*-lect of the one who knows, the talk of the culture from which his ideas, from which he himself came from. In that sense, the light of knowledge is always behind us, casting our own long shadow over the object of investigation. That is inevitable, and it is of critical importance to be aware of that condition. Rather than engaging in futile efforts to defeat reality, it is better to acknowledge our presence within the scene we research.

Each shadow is plural. Each shadow is composed of variously shaded knowledges. I want to believe that the shadow of my own knowledge is closer to that of Renoir, than to that of De Chirico (but, if possible, still geometrically relevant). But, that is only my hope. My own plural shadow is composite, distorted, appropriated – as much as it distorts and appropriates the phenomena at which it falls. That is the case with shadows each of us, and all of them, cast.
(Knowing that, Michelangelo Buonarotti asked Vitoria Colona to tell him if her beauty was so immense indeed, or it existed only in himself, magnified thousands of times by the power of his deep emotions.)

The Mirror – A Note Responsibility

The Mirror *about an ethical imperative in approaching fragile qualities of* the Other.

Behind each human endeavour, there should be a humane cause. The times of crisis have to widen that call, to include environmental and cultural concerns.

Is that a utopian thought? Yes, it absolutely is!

There is an urgent need for a paradigm shift. The extent of the current crises makes utopia indispensable. Adaptation, mitigation and similar popular(ised) soft practices do not address the causes of the crisis, and they can not succeed. Globalisation, as simple and as simplistic as it was designed to be, efficiently concentrates the power and wealth, while multiplicity and diversity of human and environmental realities remain complex - and hopelessly powerless. Since the collapse of financial centres, the questioning of the ruling paradigm reaches beyond the green agendas, where it once originated. Those questions (should) have the power shake the ruling paradigm.

Despite its heights, the XX century was a century of wars and destruction – the wars on other humans, on their cultures, on other species, on environment. A radical departure from that needs to be based on the strong ethical basis of the, at once one and diverse, humankind.

Knowledge conquers. In his *Writings of History*, de Certeau reads an allegorical etching of Vespucci by Jan Van der Straet. The explorer approaches a naked native women and "after a moment of stupor, on the threshold dotted with colonnades of trees, the conqueror will write the body of the other and trace there his own history". The Vespuccis of this world should see themselves doing that.

The naked researcher needs a mirror, to see herself in the context of the action. (It takes courage to see one's own reflection.)

But, we also need to remember that mirrors can never be trusted. The mirage they produce warrants critical eye, as they shift left for right and history shows how common and dangerous such shifts are. We need Ferrarotti's strength to "prefer not to understand, rather than to colour and imprison the object of analysis with conceptions that are, in the final analysis, preconceptions".

***

The cover page of this book is a mirror.
It invites the reader to see her/himself in the places presented by the visual essays of this book.

These Visual Essays ...

ヴィジュアルエッセイについて…

Darko Radović | ダルコ・ラドヴィッチ

Following *Tokyo dérive* in 2012 (Radović, 2013), *Mn'M* co+labo radović team has conducted a number of brief, intense *Asia dérive* experiment in Hong Kong, Bangkok and Singapore, in collaboration with *Mn'M* local partners from Chinese University of Hong Kong – Tieben and Chung teams), Chulalongkorn and Silpakorn Universities, Bangkok Issarathumnoon and Kasemsook teams), and National University of Singapore (Heng team). The idea was to further enrich and diversify the results of our investigations of urban intensities, by adding the component of cross-cultural dialogue and emphasising the value of exchange between the local and visiting teams, between the variously rooted home knowledges and visitors levels of liberating ignorance. As in Raoul Vaneigem's once reflected on his drifts with Guy Debord, our derives also provided some "exceptional moments, combining theoretical speculation, sentient intelligence, the critical analysis of beings and places, and the pleasure of cheerful drinking" (Wark, 2013). (In terms of the last item on Debords and Vaneigem' the agenda, our drifts were much poorer. After all, Debord himself; confessed: "I have written much than the most people who write, but I have drunk much more than most people who drink" (Debord, 2004). We could not stand any chance there, anyway. But, we have certainly succumbed to local advice and tasted various Chinese, Thai and Indian culinary specialties, making sure that our olfactory and gustatory immersion was of equal intensity as that of the two old Situs.)

In my call to *dérive*, I asked the team members to, in an appropriately non-hierarchical fashion, take the academic *habitus* off, to bring their own sensibilities, inspirations and values in. *Mn'M* fieldworks in Hong Kong, Bangkok and Singapore were designed to let go, to drift in spatial, experiential and intellectual sense.

The idea of *dérive* was there to help the researcher attain the necessary level of freedom, to empower her/him to drift away and to, without inhibition, seek, find, capture (somehow) and to (try to) share the impressions and those little acts of knowledge acquired in the streets of Hong Kong, Bangkok and Singapore, while those were still fresh and vulnerable. The focus was on the phenomena which seem to be of critical importance for definition of particular urban qualities which we feel, but find difficult to pin down. Discussions between the local and visiting team members were of critical definition in our attempts at exhausting the discovered places and practices (Perec, 2010).

There was also a conscious effort to challenge the domineering objectification, to postpone theorizing and thinking how the research could be useful, to reach towards simple, open-minded experiencing, recording, sharing. The toughest ask which essayistic approach to research puts in front of the researcher was to make her/himself an arena, to live *this* concrete city, to let it in, at *this* concrete moment – *hic and nunc*. As such, *dérive* celebrates subjectivity, but that of a special kind – being singular plural (Nancy, 2000). That is subjectivity capable of acknowledging layered, knotted, complex realities, without the fear of not understanding. Those phenomena may, simply, need knowledges of some other kind, the knowledges of smells and sounds, of hand or, indeed, of the *total vécu*.

Each of the three projects that make *Asia dérive* started with thorough preparations by the hosts and visiting teams. The pre-*dérive* work included revisiting the city-specific *Mn'M* material, making draft *dérive* catalogues, compiling suggestions for various "must experience", "must-feel", "must-see" moments, preparation of maps and other *dérive* materials and tools. The teams were getting ready to seek specific urban qualities/intensities, practices and situations which seem to facilitate the emergence of such intensities/qualities and, in particular, the examples of interface between public and private spaces and practices. The idea was to get as ready as possible, and then – drift away from the pre-defined, whenever and wherever some puzzling sight, smell or sound hints at possible discovery. An important reminder, based on previous investigations of this kind, was that some of the subtlest, finest qualities, those which get labeled non-measurable and, thus, dismissed as unimportant, may not necessarily be within the realm of visual perception, but in those of sounds, smell, tastes or touch.

What follows are selected traces of discovery, recorded by Mn'M *dériveurs* in Hong Kong, Bangkok and Singapore in May, June and July 2013, and formulated as visual essays by co+labo Radović research assistants Milica Muminović, Ilze Paklone, Tamao Hashimoto and Rafael A. Balboa. Graphical designer Boris Kuk has created "compressions" which open each of the three essays, and Kyushu Pistol have added final presentation touches to the visual section of *Subjectivities in Investigation of the Urban: The Scream, the Shadow and the Mirror*. Here we offer to the reader an armchair dérive, both as an immersion into someone-else's HK, BKK, SG and as hints that may help intuit some of the qualities which lay captured, but uncovered within the intense verticality of Hong Kong, gentle and rough flows of Bangkok and tamed and untamed colours of Singapore. A commited reader might also find some (inaudible but precious) sounds, tastes and smells of those places – which would, then, remain only her/his own. Let's see how it goes.

While drifting through the pages that follow, I advice the reader to place a small mirror on the page.
That is an invitation. See yourself there. Feel yourself there. Think yourself there. Think! Get reflected. Reflect!

D.R

References

Andermatt Conley, V. 1992. *Hélène Cixous*. Toronto: University of Toronto Press.
Bachelard, G. 1996. *The Poetics of Reverie*. Boston: Beacon Press.
Barthes, R. 1982. *Empire of Signs*. New York: Hill and Wang.
Bellos, D. 1993. *Georges Perec, A life in words*. Boston: David R. Godine Publisher.
Coverley, M. 2006. *Psychogeography*. Harpenden: Pocket Essentials.
Debord, G. 1998. *Comments on the Society of the Spectacle*. London: Verso.
Debord, G. 1994. *Society of the Spectacle*. New York: Zone Books.
Debord, G. 1959. Theory of dérive. In *Situationist International*, anthology. ed. Knabb, Berkeley: Bureau of Public Secrets.
Debord, G. 1998; first ed. 1988. *Comments on the Society of the Spectacle*. London, Brooklyn: Verso.
Debord, G. 2004. *Panegyric*. London: Verso.
Deleuze, G., and Guattari, F. 1987. *A Thousand Plateaus*. Minneapolis: University of Minnesota Press.
Derrida, J. 2005. *On Touching – Jean-Luc Nancy*. Stanford: Stanford University Press.
Derrida, J. 2006. *The Politics of Friendship*. London: Verso.
Haraway, D. 1991. *Simians, Cyborgs, and Women: The Reinvention of Nature*. Oxford: Routeldge.
Harvey, D. 1990. *The Condition of Post-Modernity*. Molden: Blackwell Publishing.
Highmore, B. 2006. *Michel de Certeau, Analysing Culture*. London: Verso.
Hillis-Miller, J. 2001. *Others*. Princeton and Oxford: Princeton University Press.
Howes, D. 2005. *Empire of the Senses*. ed. Howes, Oxford: Berg.
Jacobs, J. 1965. *The Death and Life of the Great American City*. New York: Random House.
Law, J. 2004. *After Method: Mess in social science research*. Abingdon, New York: Routledge.
Lefebvre, H. 1996. *Writings on Cities*. Cambridge, Ma.: Blackwell.
Nancy, J-L. 2000. *Being Singular Plural*. Stanford: Stanford University Press.
Merleau-Ponty, M. 2012. *Phenomenology of Perception*. Abingdon: Routledge.
Mitchell, S. 2012. *Unsimple Truths – Science, complexity and policy*, Chicago: University of Chicago Press
MOMA, 2013, http://www.moma.org/visit/calendar/exhibitions/1330 (visited 13.1.2014)
Pallasmaa, Y. 2005. *The Eyes of the Skin*. Chichester: John Wiley & Sons Ltd.
Radović, D. 2008. *eco-urbanity*. Abingdon, New York: Routledge.
Radović, D. 2012. The greatness of small. In *FUTURE–ASIAN-SPACE: Projecting the urban space of new East Asia*. eds. Hee, Viray, Boontharm, Singapore: NUS Press, pp. 159–170.
Perec, G. 2010. *An Attempt at Exhausting a Place in Paris*, Cambridge, M.A.: Wakefield Press.
Radović, D. 2013. *Intensities in Ten Cities*. Tokyo: flick studio.
Radović, D. 2013a. *Tokyo Dérive – In Search of Urban Intensities*. Tokyo: flick studio.
Radović, D., with Boontharm, Kuma, Grgić, eds. 2012. *The Split Case*. Tokyo: flick studio.
Roberts, L. 2005. The death of sensuous chemist. In *Empire of the Senses*, ed. Howes, Oxford: Berg, pp. 106-127.
Serres, M. 2008. *The Five Senses*. London: Continuum Publishing.
Sheringham, M. 2006. *Everyday Life, Theories and Practices from Surrealism to the Present*. Oxford: Oxford University Press.
Schumacher, E.F. 1973. *Small is Beautiful: economy as if people mattered,*. London: Harper Collins Publishers.
Spengler, O. 1928. *The Decline of the West, vol. 2*. New York: Alfred A. Knopf Inc.
Wark, McKenzie. 2013. *The Spectacle of Disintegration*. London: Verso.
West-Pavlov, R. 2009. *Space in Theory: Kristeva, Foucault, Deleuze*. Amsterdam, New York: Rodopi.
Žižek, S. 2005. Against human rights. In *New Left Review 34*, pp. 115–131.

HENDRIK TIEBEN
ヘンドリック・ティーベン

Drifter_

Situation_ Hong Kong is often regarded as a model of a vertical city in Asia. On Hong Kong Island and in Kowloon a 3d circulation network emerged, facilitated by steep topography and land scarcity and links now various urban infrastructures, topographical levels and multiple programs. This network provides a highly efficient connectivity at all weather conditions.

The derive "3d Hong Kong" was organized as one day drift through this three-dimensional network exploring how different public-private space interfaces organize accessibility, social activities and urban experiences along vertical sections. The derive was set in motion by a set of simple rules encouraging vertical movements through a range of different building types. It started at Hong Kong Station and the International Finance Center, currently highest tower on Hong Kong Island and the center of Hong Kong's financial industry. From there it drifted to the area around the Central Escalator system passing street eateries and informal occupations.

Along the way different connections and boundaries between public and private spaces were captured. With a GPS recorder, the direction and speed of the vertical and horizontal movements were traced as well as the locations of photos. During the derive particular attention was given to the way, how the different arrangements of vertical interfaces facilitate or discourage activities.

CENTRAL HK
香港中環

Dérive in CENTRAL HONGKONG 028 | 029

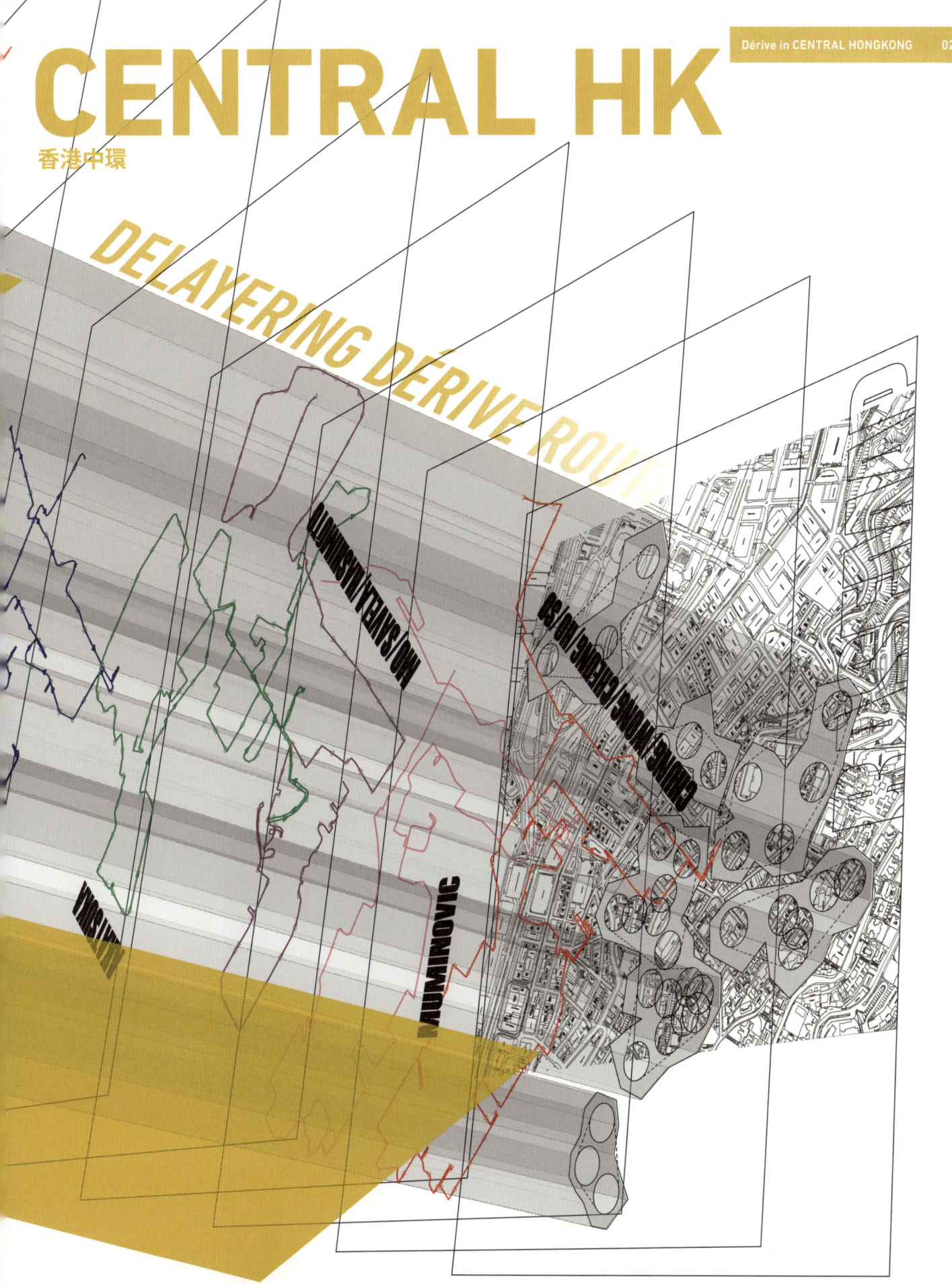

CENTRAL HK

香港中環

SYMBOLIC VERTICALITY **HIGH RISE**

HUMIDITY IS IN THE AIR

Drifter_
MIKA SAVELA
ミカ・サーベラ

Situation_
The urban environment of today has many qualities, formed and modeled by politics, activities, control, architecture, planning and design. They occur in variety of frameworks, some obvious, some invisible, and unmeasurable. But for person drifting through a city, these occurrences might become visible in a context other than what has been originally proposed. For an outsider, an observer, seeing beyond the meaning or purpose comes almost natural at the beginning. Without the knowledge of the rules or guidelines, any scene can be seen as it visually or functionally appears–with its benefits and obstacles.

Drifter_
TAMAO HASHIMOTO

Drifter_
HENDRIK TIEBEN
ヘンドリック・ティーベン

Hendric: Red   Tamao: Purpule   Tamao and Mika: Purpule

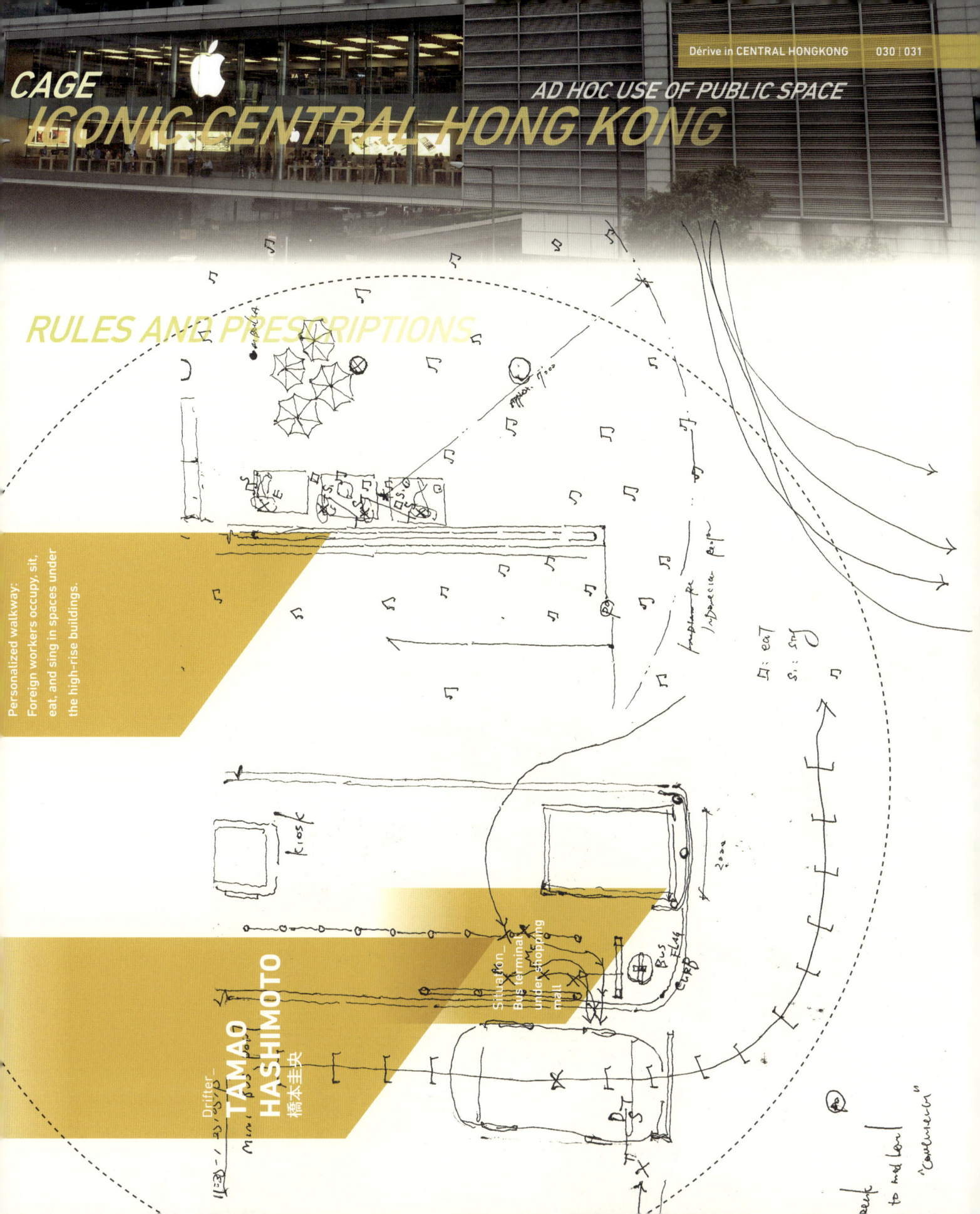

Dérive in CENTRAL HONGKONG 030 | 031

CAGE
ICONIC CENTRAL HONG KONG

AD HOC USE OF PUBLIC SPACE

RULES AND PRESCRIPTIONS

Personalized walkway: Foreign workers occupy, sit, eat, and sing in spaces under the high-rise buildings.

Drifter_Mini Bus stop_ **TAMAO HASHIMOTO** 橋本圭央

CENTRAL HK
香港中環

MOVEMENTS

Drifter_
TAMAO HASHIMOTO
橋本圭央

Situation_
Central Escalator for sightseeing photo and exhibition purpose.

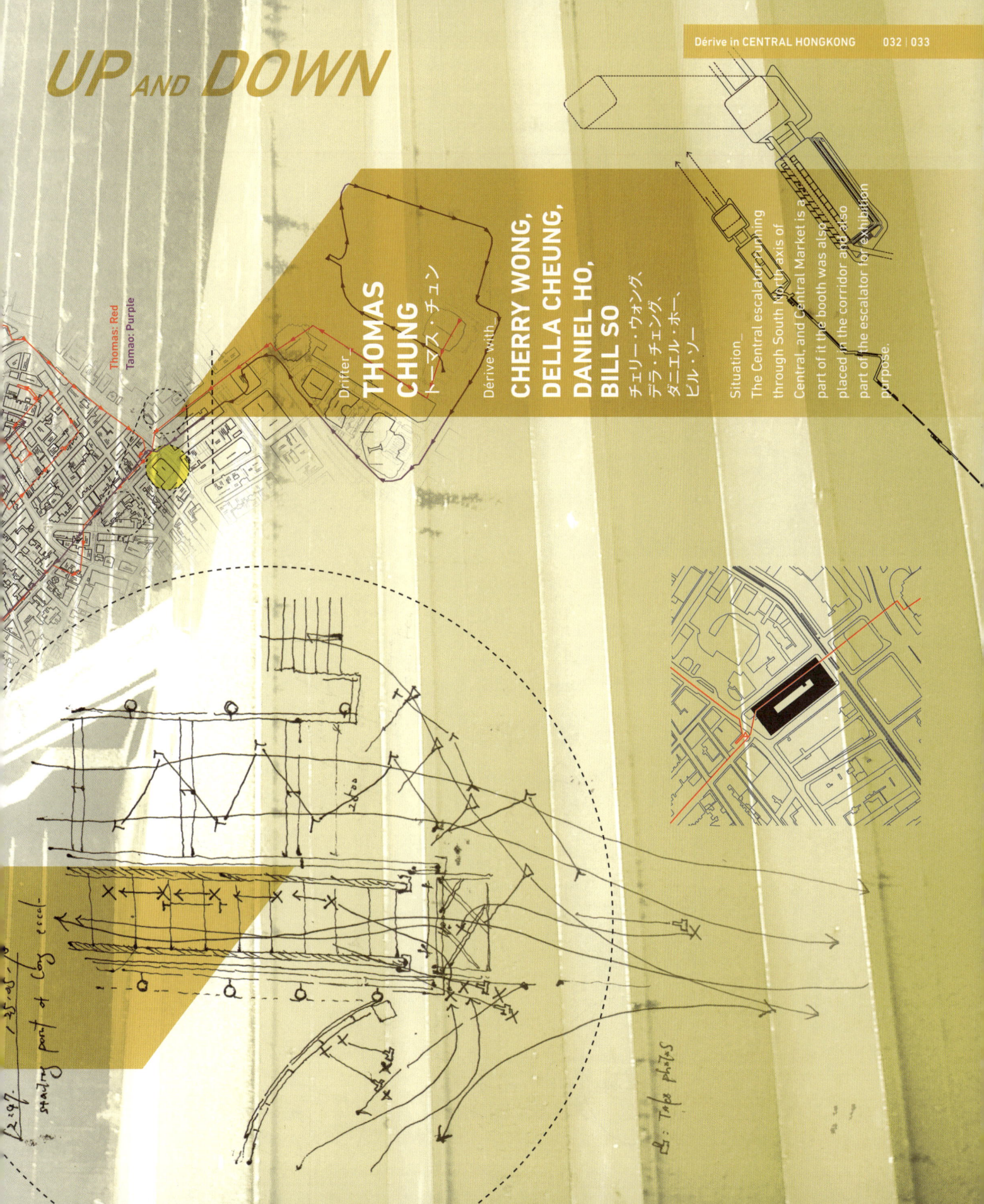

CENTRAL HK
香港中環

CONTROLLED
FENCED
EMPTYNESS
LIVING IN SECURED WORLD

Ken: Green
Mika: Purple

Dérive in CENTRAL HONGKONG 034 | 035

Void, Emptiness, Irony, Abandoned Luxury, **Abstraction**

ADVERTISEMENT

Drifter_
MIKA SAVELA
ミカ・サーベラ

Situation_
Spaces of futile design labour, where the imagery becomes a display of voids, emptiness, irony, abandoned luxury, abstraction. The observer becomes the displayer and revealer of mismatched intentions and the realities of the urban space.

Drifter_
KEN AKATSUKA
赤塚 健

GREEN, HUMID, FRESH MOIST
INTENSITY OF EXCHANGE

Dérive in CENTRAL HONGKONG 036 | 037

Ken: Green
Milica: Pink

Drifter_
KEN AKATSUKA
赤塚 健

Situation_
Green makes people relaxed whether it is natural or artificial. Small greens in public space would lead the private activities. City green creates the area of boundary between public and private with making the scenes of daily life of the urban dwellers. It would allow us to understand the lives of people and describe the city or social structures by focusing on the urban green. Atriums of super high-rise buildings, public stairs on the hill, gated park in the city area, or the bottom of the pencil residential towers... Drifting the over-floated city and seeking the places, opening one's mind and creating active lives.

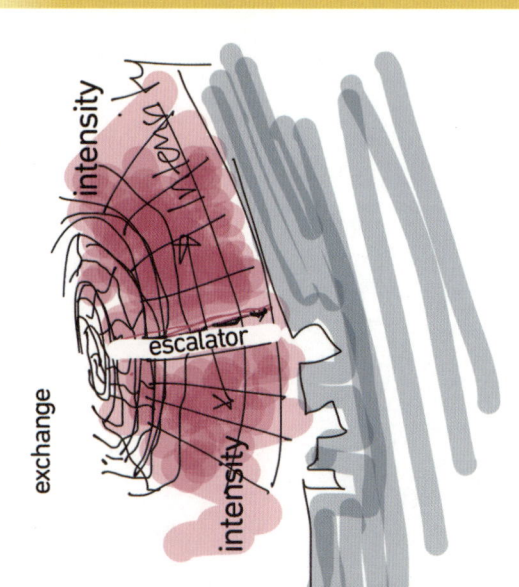

CENTRAL HK
香港中環

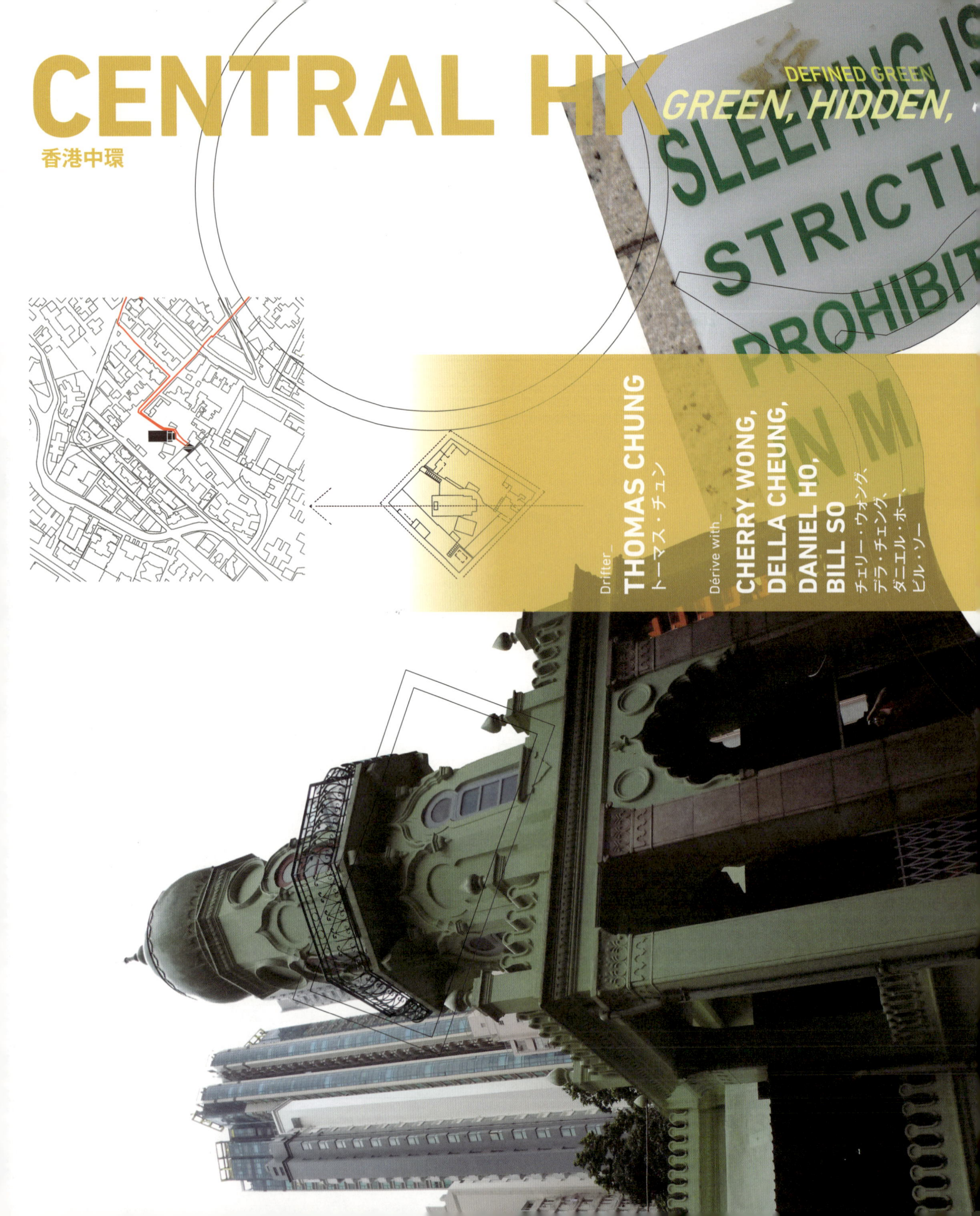

DEFINED GREEN
GREEN, HIDDEN,

Drifter_ **THOMAS CHUNG**
トーマス・チュン

Dérive with_ **CHERRY WONG, DELLA CHEUNG, DANIEL HO, BILL SO**
チェリー・ウォング、
デラ・チェング、
ダニエル・ホー、
ビル・ソー

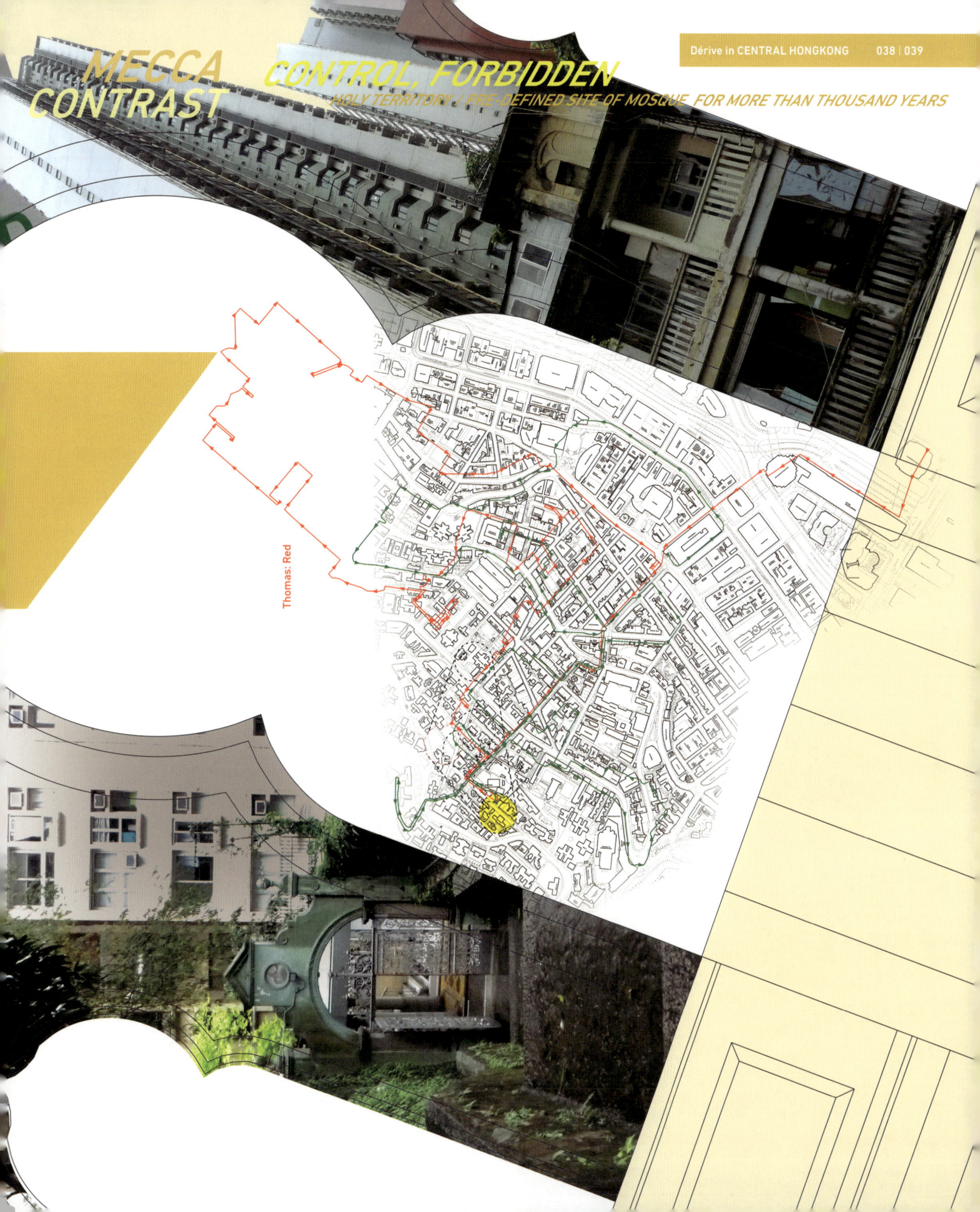

MECCA CONTRAST — **CONTROL, FORBIDDEN**
HOLY TERRITORY / PRE-DEFINED SITE OF MOSQUE FOR MORE THAN THOUSAND YEARS

Dérive in CENTRAL HONGKONG 038 | 039

Thomas: Red

CENTRAL HK

香港中環

JUXTAPOSITION OF CONTRAST
URBAN REQUALIFICATION
AMALGAM

RESILIENCE

Drifter_
THOMAS CHUNG
トーマス・チェン

Dérive with_
**CHERRY WONG,
DELLA CHEUNG,
DANIEL HO,
BILL SO**
チェリー・ウォング、
デラ・チェング、
ダニエル・ホー、
ビル・ソー

Thomas: Red
Davisi: Blue

REINTERPRETATION OF THE EXISTING
OF CREATIVITY AND REUSE

Dérive in CENTRAL HONGKONG   040 | 041

Drifter_
DAVISI BOONTHARM
ダヴィシー・ブンタム

Situation_
SOHO to POHO the clash of gentrification and resilience

CENTRAL HK
香港中環

ADHOC USE OF OPEN SPACE
OPEN

Drifter_
KEN AKATSUKA
赤塚 健

Ken: Green
Mika: Purple

Dérive in CENTRAL HONGKONG 042 | 043

WHETHER ON GROUND LEVEL OR ON THE ROOFTOPS,
EMPTY GREEN SPACES ARE USED BOTH IN ADHOC AND CONTROLLED MANNERS
ADOHOC, VERTICARITY, CONTROL

Drifter_
MIKA SAVELA
ミカ・サーベラ

Situation_
OPPORTUNISTIC. POSITIONS. Spaces where verticality becomes observable as oddly positioned, absurd, irrational or opportunistic activities. Whether on ground level or on the rooftops, spaces are used both in adhoc and controlled manners.

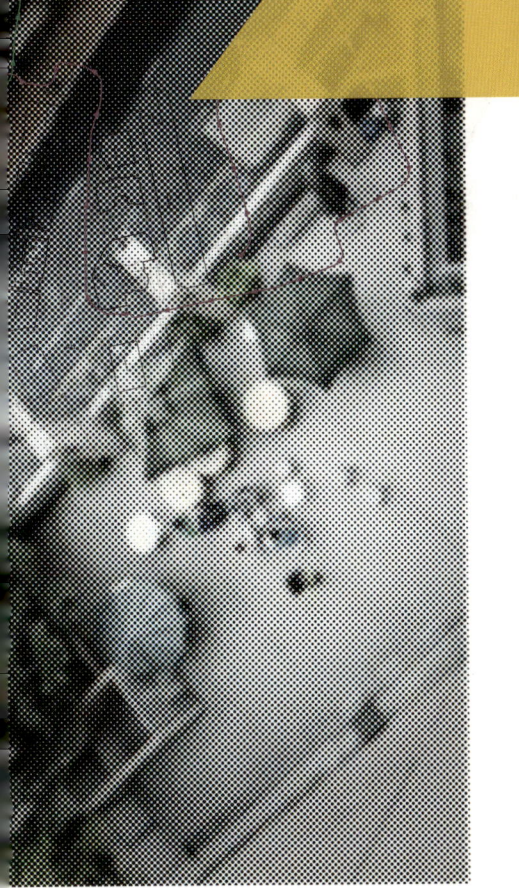

MONG KOK
モンコック

Thomas: Red
Milica: Pink
Tamao: Purple

Drifter_
THOMAS CHUNG
トーマス・チュン

Dérive with_
**CHERRY WONG,
DELLA CHEUNG,
DANIEL HO,
BILL SO**
チェリー・ウォング、
デラ・チェュング、
ダニエル・ホー、
ビル・ソー

Dérive in MONG KOK 044 | 045

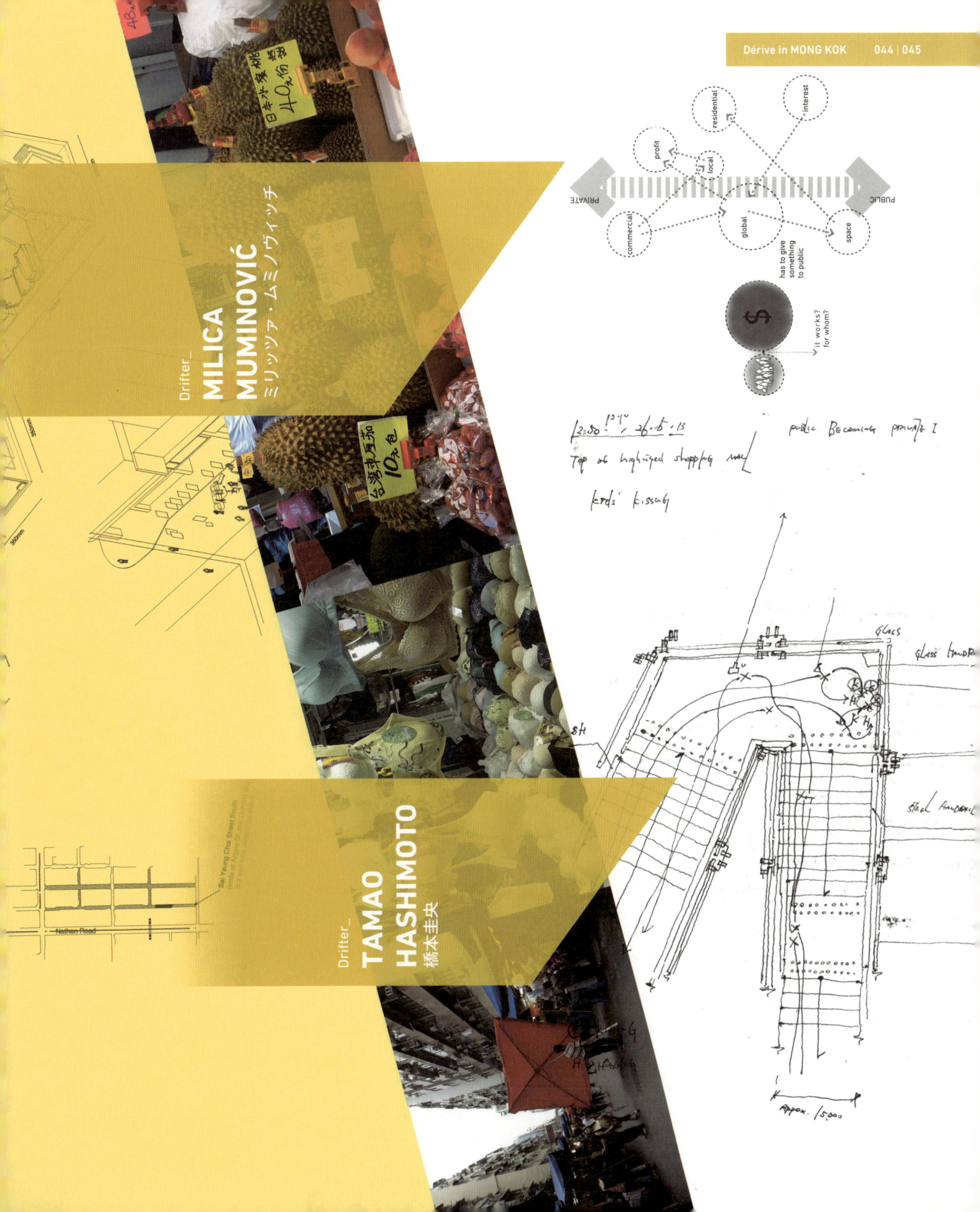

Drifter_
MILICA MUMINOVIĆ
ミリッツァ・ムミノヴィッチ

Drifter_
TAMAO HASHIMOTO
橋本圭央

public Becoming private I
Top of highrised shopping mall
kids' kissing

MONG KOK
モンコック

HUMIDITY

Drifter_
DARKO RADOVIĆ
ダルコ・ラドヴィッチ

IS IN THE AIR

Dérive in MONG KOK 046 | 047

MONG KOK
モンコック

RIGHT TO THE

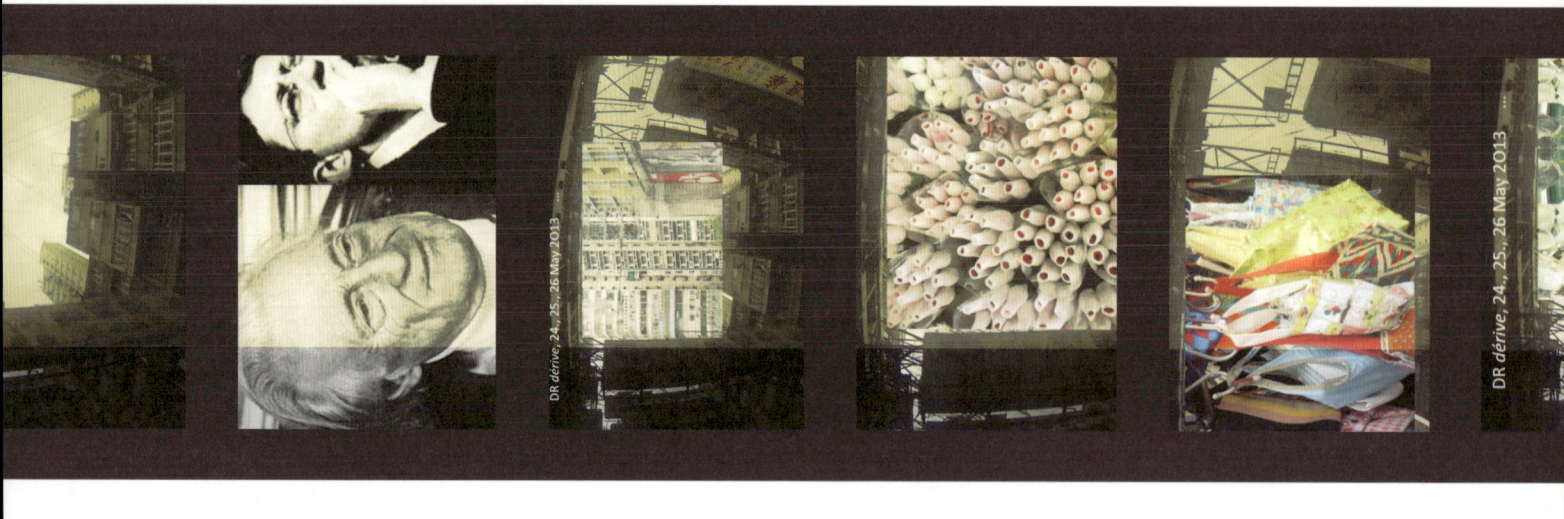

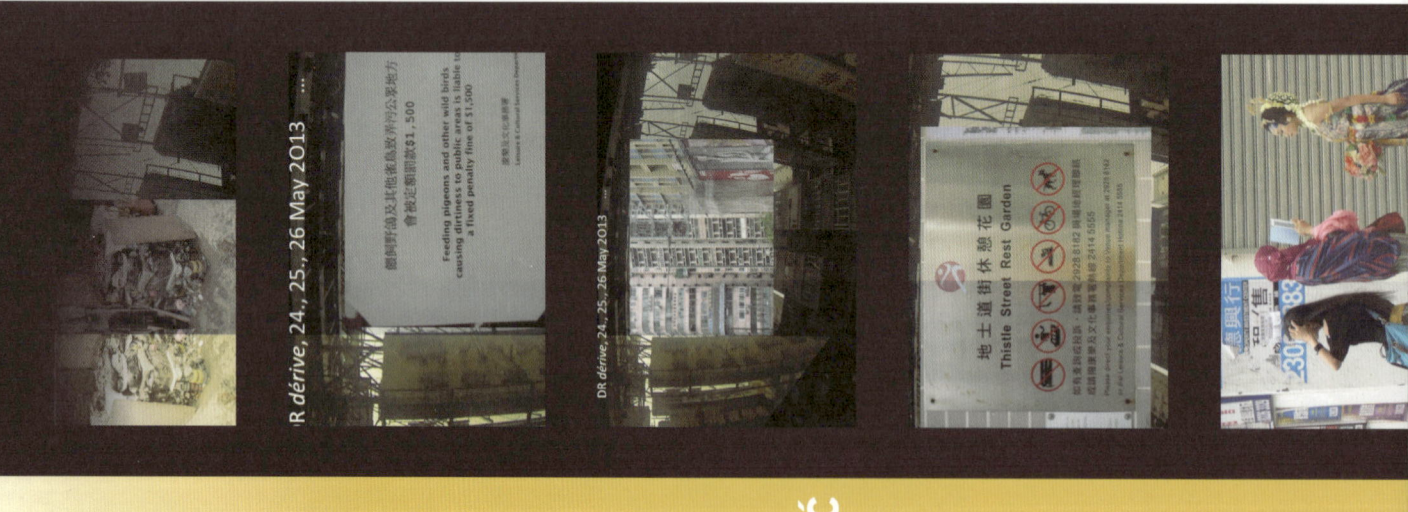

Drifter_ **DARKO RADOVIĆ**
ダルコ・ラドヴィッチ

CITY EXCLUSION
DOMINATION OF PRIVATE ESCAPE

Dérive in MONG KOK 048 | 049

BKK

Dérive in BANGKOK／バンコク

Bunglumphu / Sukhumvit

VISUAL ESSAY by Milica Muminović／ミリッツァ・ムミノヴィッチ

based on the work by: Akatsuka, Boontharm, Hashimoto, Issarathumnoon, Kanittaprasert, Karnchanaporn, Kasemsook, Kunajitpimol, Laithavewat, Muminović, Sanguandekul, Seneewong Na Ayudhaya, Teangtawat,Thittayabhorn, Tinrat and Wittayalertanya

fieldwork by: co+labo radović team (Darko Radović, Davisi Boontharm, Milica Mimonović, Tamao Hashimoto, Ken Akatsuka, Yoshiaki Kato), Chulalongkorn University (Kanyapat Seneewong Na Ayudhaya, Thittayabhorn Mitudom, Parima Amnuaywattana, Prat Tinrat, Sasinee Teangtawat, Siwat Wittayalertanya, Wimonrart Issarathumnoon) and Silpakorn University (Apiradee Kasemsook, Dherapat Sanguandekul, Methi Laithavewat, Monsicha Kanittaprasert, Nuttinee Karnchanaporn, Thiti Kunajitpimol)

Bangkok intensity – urban compression by Boris Kuk

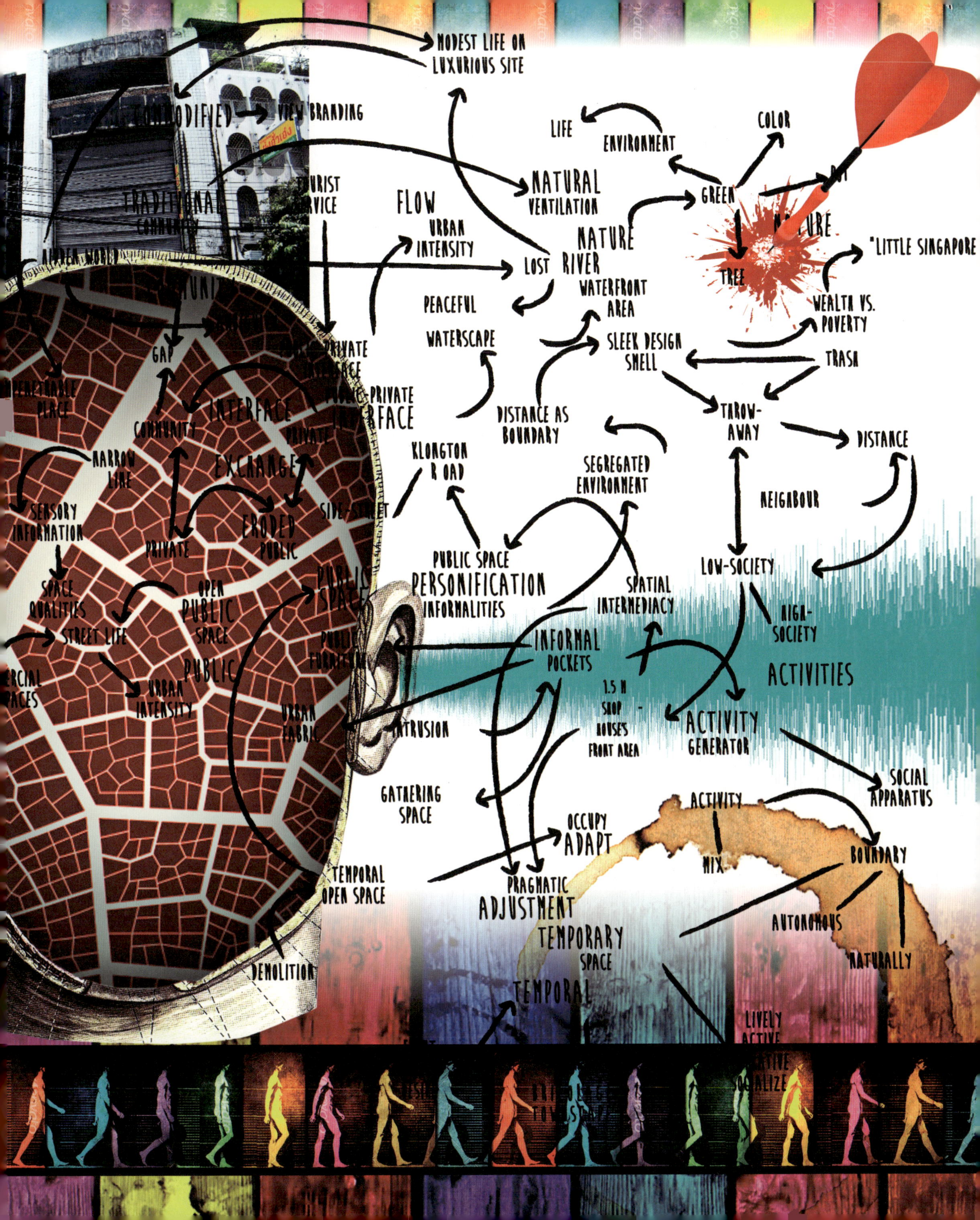

MODEST LIFE ON LUXURIOUS SITE
COMMODIFIED
VIEW BRANDING
TRADITIONAL COMMUNITY
TOURIST SERVICE
FLOW
URBAN INTENSITY
NATURAL VENTILATION
LIFE
ENVIRONMENT
COLOR
GREEN
NATURE
LOST
RIVER
TREE
"LITTLE SINGAPORE
PEACEFUL
WATERFRONT AREA
WEALTH VS. POVERTY
WATERSCAPE
SLEEK DESIGN SHELL
TRASH
GAP
PUBLIC-PRIVATE INTERFACE
INTERFACE
COMMUNITY
PRIVATE
DISTANCE AS BOUNDARY
THROW-AWAY
DISTANCE
NARROW LINE
EXCHANGE
KLONGTON R OAD
SEGREGATED ENVIRONMENT
SENSORY INFORMATION
SIDE-STREET
NEIGHBOUR
ERODED PUBLIC
PRIVATE
PUBLIC SPACE
LOW-SOCIETY
SPACE QUALITIES
OPEN PUBLIC SPACE
PUBLIC SPACE
PERSONIFICATION
INFORMALITIES
SPATIAL INTERMEDIACY
STREET LIFE
HIGH-SOCIETY
INFORMAL POCKETS
PUBLIC
ACTIVITIES
URBAN INTENSITY
1.5 M SHOP - HOUSE'S FRONT AREA
URBAN FABRIC
INTRUSION
ACTIVITY GENERATOR
SOCIAL APPARATUS
GATHERING SPACE
ACTIVITY
OCCUPY
ADAPT
MIX
BOUNDARY
TEMPORAL OPEN SPACE
PRAGMATIC ADJUSTMENT
AUTONOMOUS
TEMPORARY SPACE
NATURALLY
DEMOLITION
TEMPORAL
LIVELY ACTIVE

052 Dérive in SUKHUMVIT

Boundaries Defined by Acivities
アクティビティによる境界線

Dérive in_
Sukhumvit
スクンビット

Sukhumvit 71 Road-street view

temporal

Sukhumvit Rd.

Drifter_
Monsicha Kanittaprasert
モンシシャー・カンタープラサー

Row houses elevation

Row houses Elevation

people...
sitting
relaxing
talking
sleeping
in public space...
the boundaries are not defined by the architectural elements...they are dissolved through **leaking activities**....

bathroom by the canal

kitchen by the canal

Banglumphu laundry
バンランプーのランドリー

Dérive in_
Banglumphu
バンランプー

Drifter_
Prat Tinrat
プラット・ティナット

Chaopraya River

Wat Sampraya

Wat Sangwet

Banglumphu canal

End

Prasumaru Fortress

Prasumaru RD.

Side-street Personification:
Spatial intermediacy of the Bangkok Shophouses

典型的サイドストリート：
ショップハウスがおよぼす空間的媒介

Dérive in Sukhumvit
スクンビット

Drifter Apiradee Kasemsook and Nuttinee Karnchanaporn
アピラディ・カセンスーク、ナッティニ・カルンチャポン

Drifting along Klongton Road's footpath-scape, we encountered an interesting variation of spatial **personification** of the fronts of these shophouses. Ministry of Interior's building regulation forbids a structural construction within a 1.50 meter depth area from the front of every shophouse. Although the 1.50 meter depth area belongs to the shophouse owner, the owner has to follow the set-back **regulation**. The negotiation between what belongs to the private realm of living and what needs to obey the public law takes place here in the 1.50 meter depth area from the front of shophouse.

Each shophouse owner has his/her own **technique** to reclaim the set-back front, some of them lend or rent out the space for street vendors.

A pragmatic adjustment or adaptation displays a genius negotiation for the spatial intermediacy between the private, the public, or the domesticity, the **urbanity**.

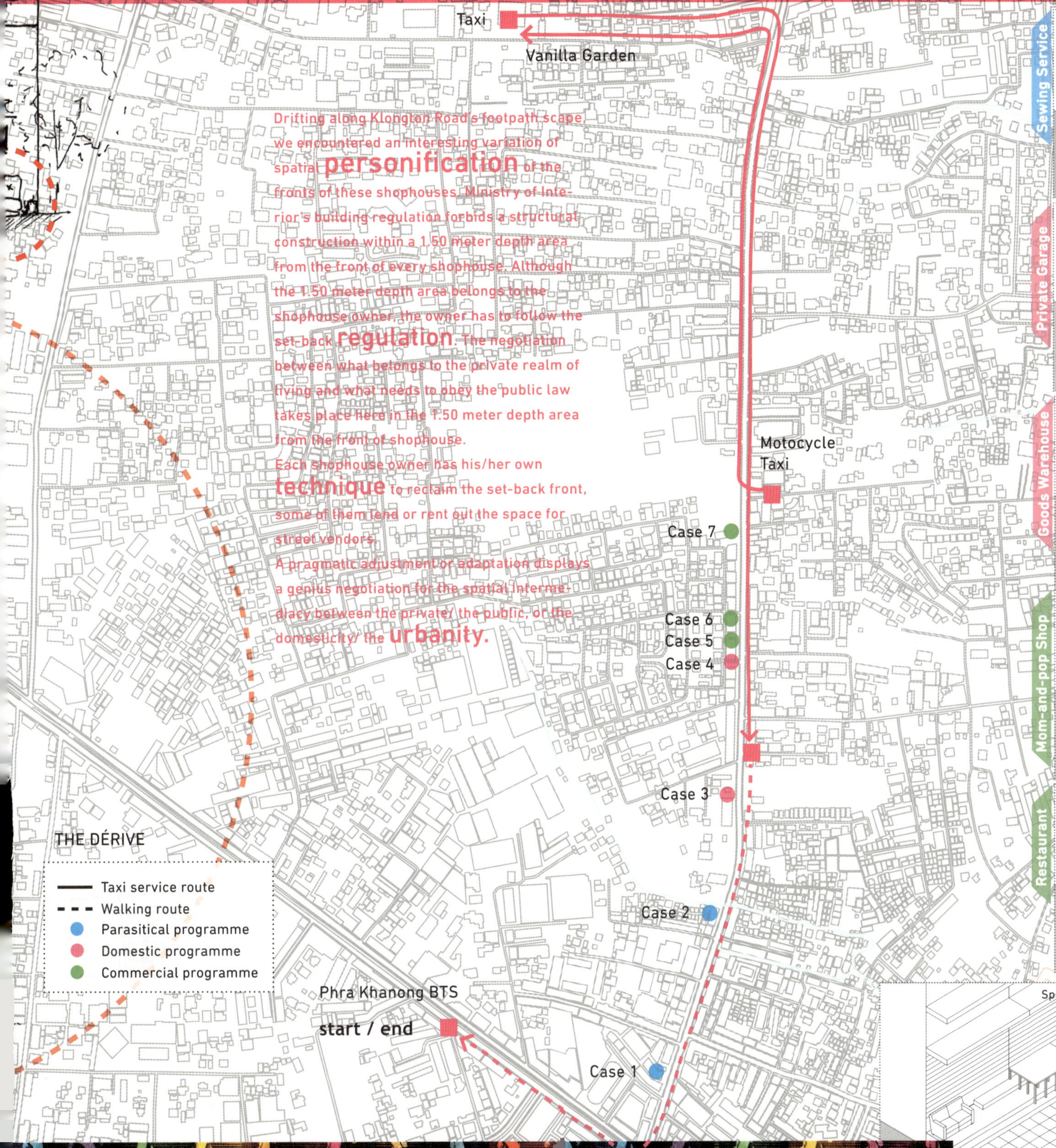

THE DÉRIVE

- ▬ Taxi service route
- ┄ Walking route
- ● Parasitical programme
- ● Domestic programme
- ● Commercial programme

Street Vendor — borrowing-seats
Sewing Service — lending space
Private Garage — folding ramps
Goods Warehouse — stairs/slopes
Mom-and-pop Shop — extending shelves
Restaurant — claiming footpath
— extending business

Spatial Intermediacy 1–7: closed / setting / operating

056 | 057 | 058   Dérive in SUKHUMVIT, BUNGLUMPHU

Super-Public-Space
スーパー・パブリック・スペース

Dérive in_ **Sukhumvit** スクンビット

Drifter_ **Thiti Kunajitpimol** ティッティ・クナッチピモン

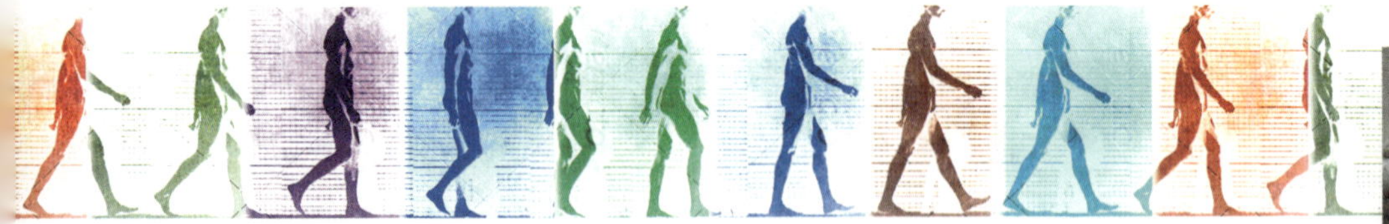

Ephemeral Space
一過性のスペース

Dérive in_ **Sukhumvit** スクンビット

Drifter_ **Methi Laithavewat** メティ・ライタウィワット

space in the gap of time between demolition and reconstruction opportunity

occupy and adapt...
temporal users...
ad-hoc users....
community users...

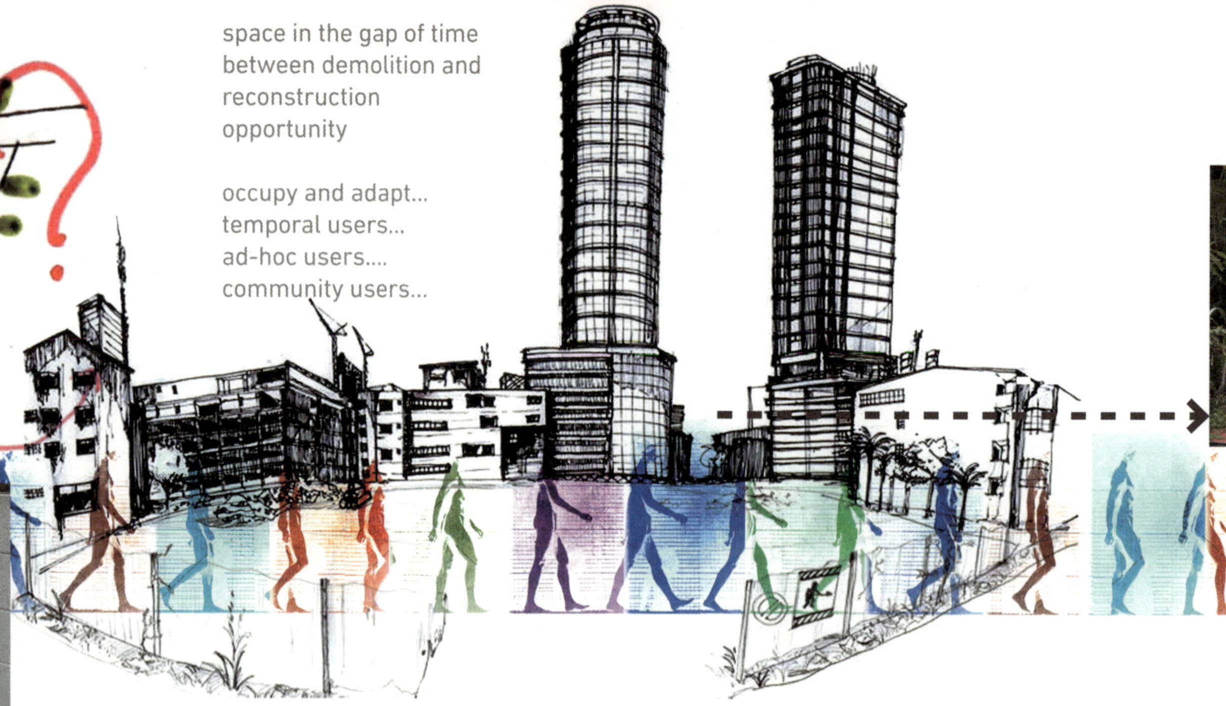

shophouse
community space/**personal** space
public space/**impersonal** space
space defined by garbage spots

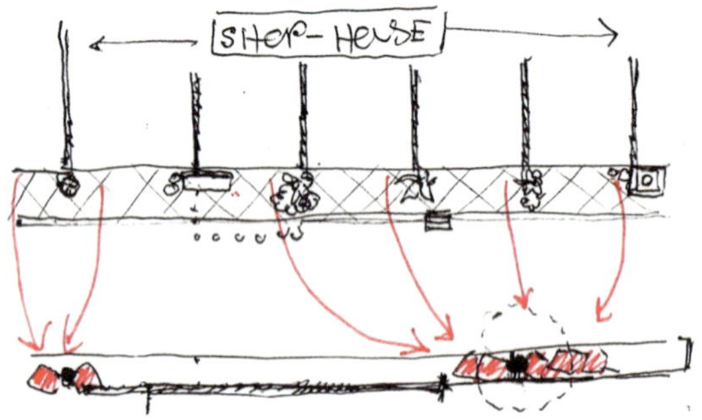

places for disposing the garbage=spaces which don't belong to anybody+ places which belong to everybody

The demolition in order to rebuild creates incidentally temporal space. This open space becomes useful urban space for neighbors. They use it as hang-out area, for sports or just as large open space.

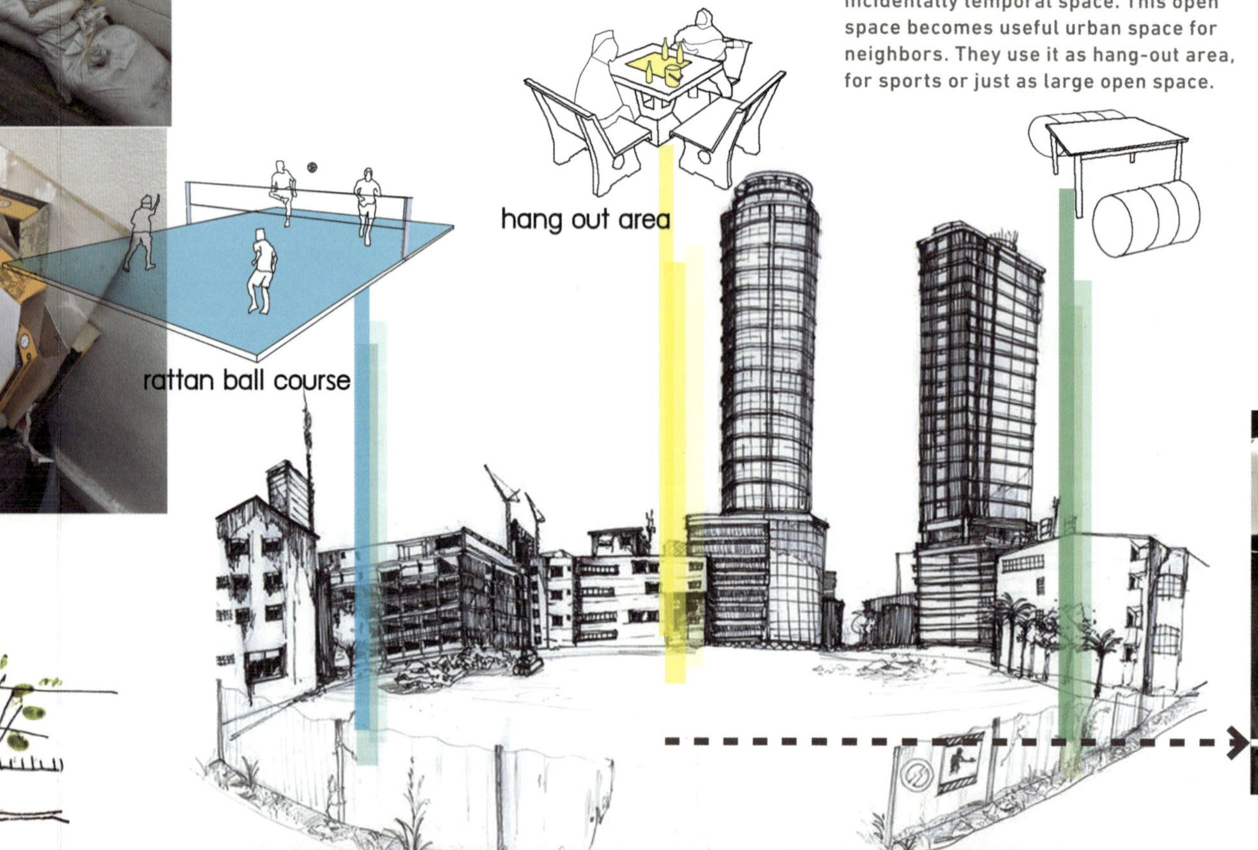

rattan ball course

hang out area

activities

garbage gravity

But, what about the quali

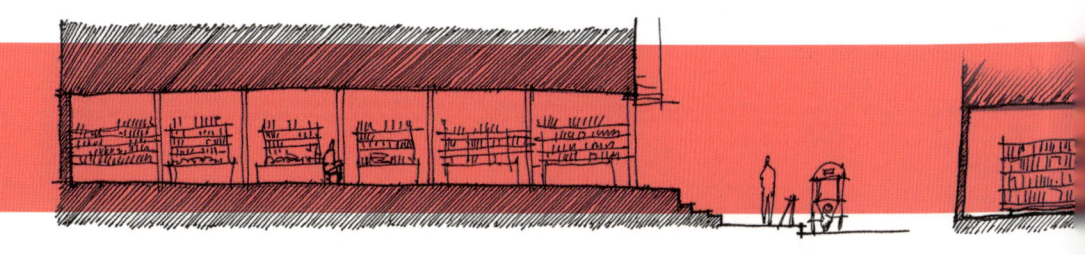

...is an L-shape shopping mall... Its spaces
...ck spaces of other shop-houses in the same

behind front view

Banglamphu Department Store was once a large-scale shopping mall...

today...the business is no longer running...

...the ground floor...rented...small stalls/upper floors...vacant...

the inner spaces were empty. I felt lonesome when I saw an escalator that has not been utilized and some upper spaces have been neglected...

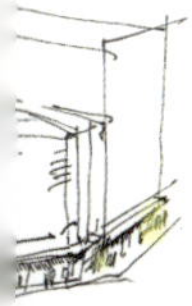

New world Department Store is used to be a famous shopping mall...it became abandoned and was issued to be demolished...

...the process of destruction of the building has been obstructed

...and the structure of the abandoned building remains ... due to the leak of building roof, some inner spaces unbelievably turned into a large pond with a number of fishes....

outdoor spaces around the building became spaces for permanent stalls and street vendors....

counters at the front of the shop settle opposite pedestrian booths

there is a way to upstairs by walking on the old escalator but the power is shut down

Dérive with
Parima Amnuaywattana
soi samsan 5

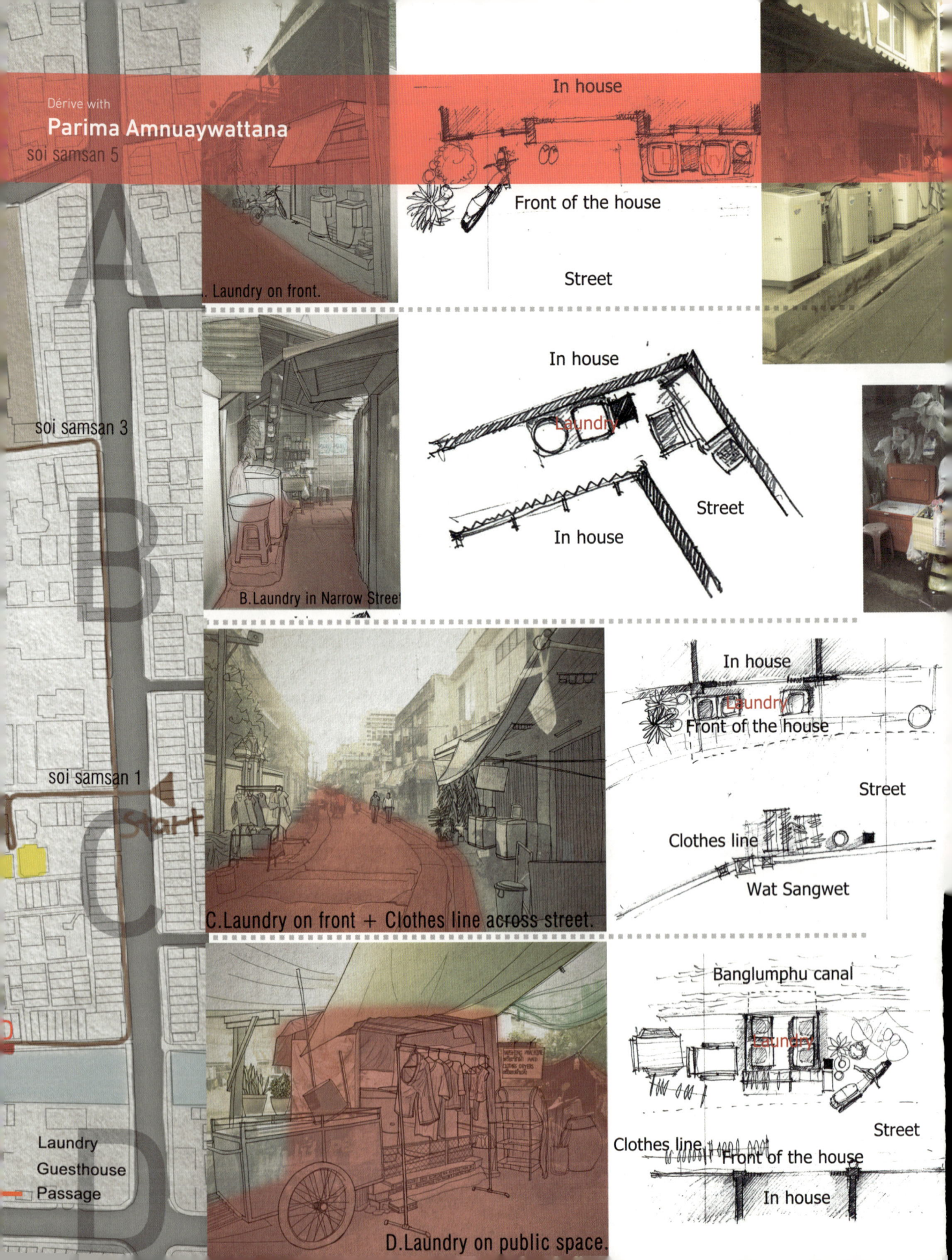

soi samsan 3

soi samsan 1

Laundry
Guesthouse
Passage

In house
Front of the house
Street

A. Laundry on front.

In house
Laundry
Street
In house

B. Laundry in Narrow Street

In house
Laundry
Front of the house
Street
Clothes line
Wat Sangwet

C. Laundry on front + Clothes line across street.

Banglumphu canal
Laundry
Clothes line
Front of the house
Street
In house

D. Laundry on public space.

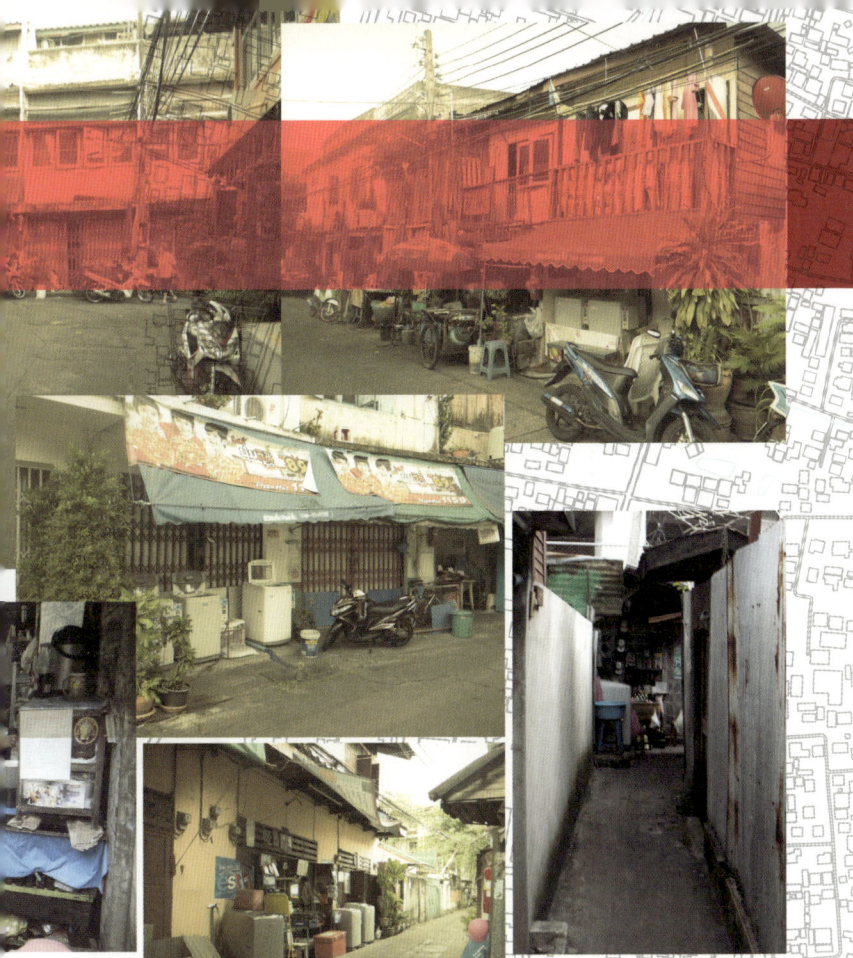

Dérive in BUNGLUMPHU, SUKHUMVIT 059 | 060 | 061

Instant Society
インスタント社会

Dérive in_
Sukhumvit
スクンビット

Drifter_
Dnerapat Sanguandekul
ティラパット・サンクンディクン

Social Machine
breeding the informal gathering
The informal contributes to the liveliness of the streets...if there were no informal activities and elements that support them the streets would be clean and easy to walk but also lethargic... How can we design the space to generate the informal social activities?

areas: Wat Sang Whet and Wat Sam Phraya Communities
a cluster of guesthouses for tourists

locals and tourists interactions through laundry service
merging private and public spaces
Types:
A. Laundry on front
B. Laundry in Narrow Street
C. Laundry on front + Clothes line across street
D. Laundry in public space

— Dérive route
● Observed "gathering space"

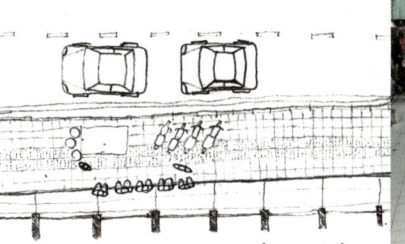

Figure A: A food stall selling everything from grilled fish to blanched vegetables on the sidewalk turns the area into a dining room with storefront steps as seats.

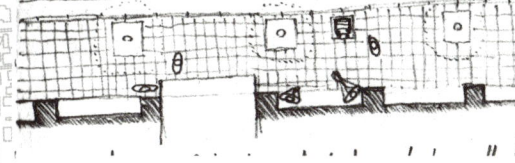

Figure B: A bus stop in front of a suburban mall. Because of long waiting time, window fronts are used as seats and large columns as leaning posts, and people sit and wait as if they were having an appointment with the doctors.

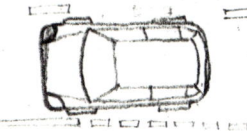

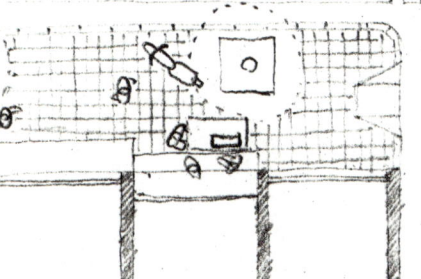

Figure C: A sewing machine placed in front of a Laundromat. The owner and her two acquaintances sit/stand in gossip as though it were their living room.

Figure D: An area slightly enclosed by a telephone booth and an overcrossing's staircase suddenly becomes a safe heaven for neighbouring children to make use as their playground, with adult supervision.

Unused Commercial Spaces
未使用のコマーシャルスペース

Dérive in_ **Banglumphu** バンランプー

Drifter_ **Siwat Wittayalertanya** シワット・ウィタカラータンヤ

Dérive with **Yoshiaki Kato, Kanyapat Seneewong Na Ayudhaya and Sasinee Teangtawat**

shopping district...

over time...large scale department stores lost the attractiveness....and the commercial function...

My dérive focused on the unused spaces of the department stores. It was aimed at illustrating typology of shopping areas and unexploited areas, particularly focusing on **urban intensity**. It was based on the main question: why the commercial spaces in a form of **informal** hawkers are more vigorous and gain more popular for the customers than the **large scale** department stores which have gradually lost their vibrancy?

urban intensity
shaped by various kinds of crowded hawkers in public areas
1. connected to commercial spaces inside the buildings
2. independent on the streets
3. mobile

counter   street vendor   booths   street food

booths outside the mall

2 storeys commercial space, first storey is fashion clothing

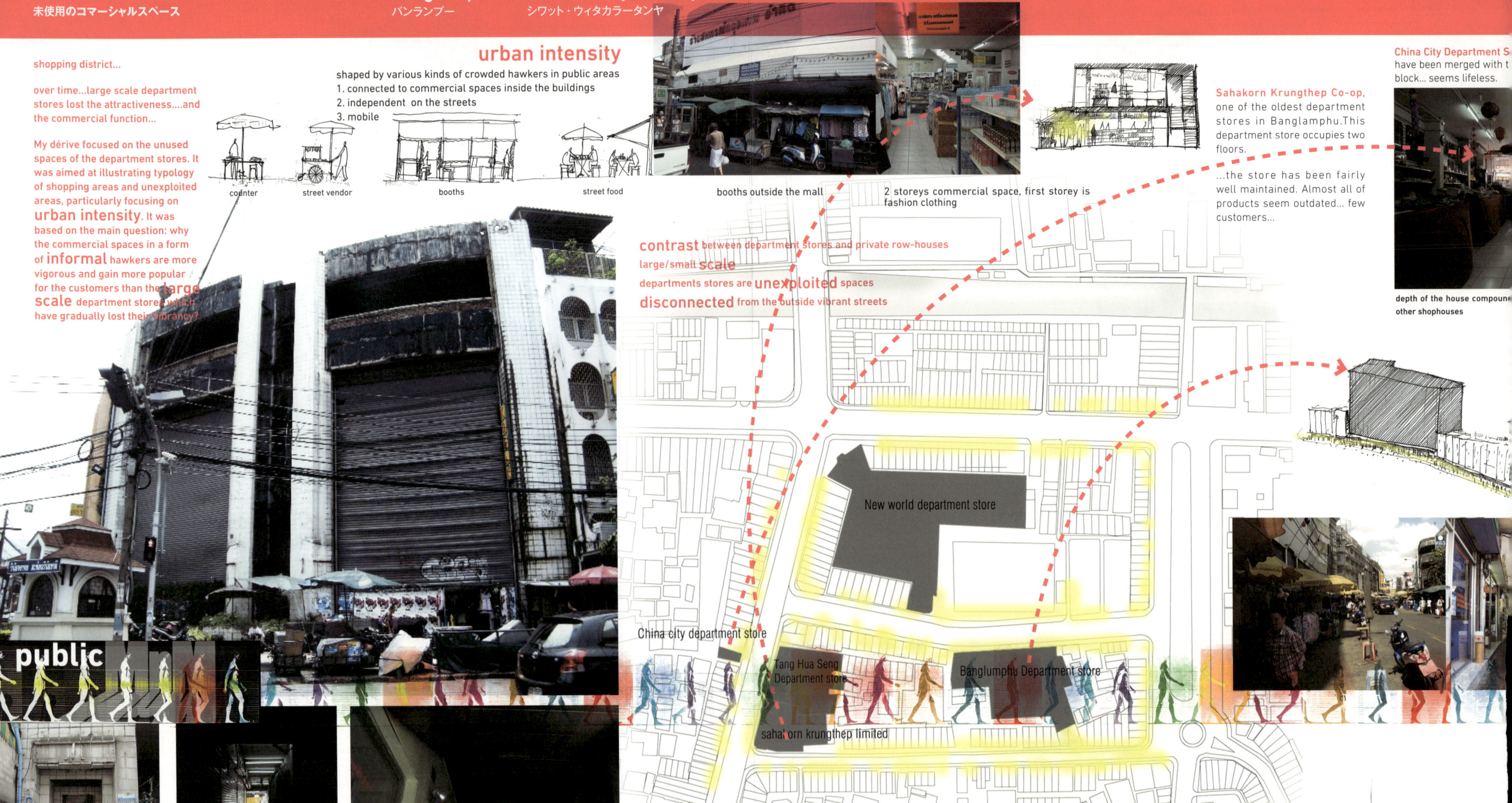

contrast between department stores and private row-houses
large/small **scale**
departments stores are **unexploited** spaces
disconnected from the outside vibrant streets

China City Department S have been merged with t block... seems lifeless.

Sahakorn Krungthep Co-op, one of the oldest department stores in Banglamphu. This department store occupies two floors.

...the store has been fairly well maintained. Almost all of products seem outdated... few customers...

depth of the house compound other shophouses

public

New world department store

China city department store

Tang Hua Seng Department store

Banglumphu Department store

sahakorn krungthep limited

= department store's commercial spaces

= booths and street venders

Dérive in BUNGLUMPHU   065

Tang Hua Seng Department Store

the first floor of the building has been used as a supermarket with numerous stalls and counter cashiers...

...few people come to shop here

...but the stalls and street vendors outside the supermarket are crowded...

Viewpoint

...the informal vendor spaces along the street persuade people to walk through...people walk slowly ...pass the narrow ways... see various kinds of goods ...

...once inside the enclosed spaces of the department stores people feel uncomfortable... low ceiling and surrounded walls..people will go inside the department stores only when they want to buy some specific things....

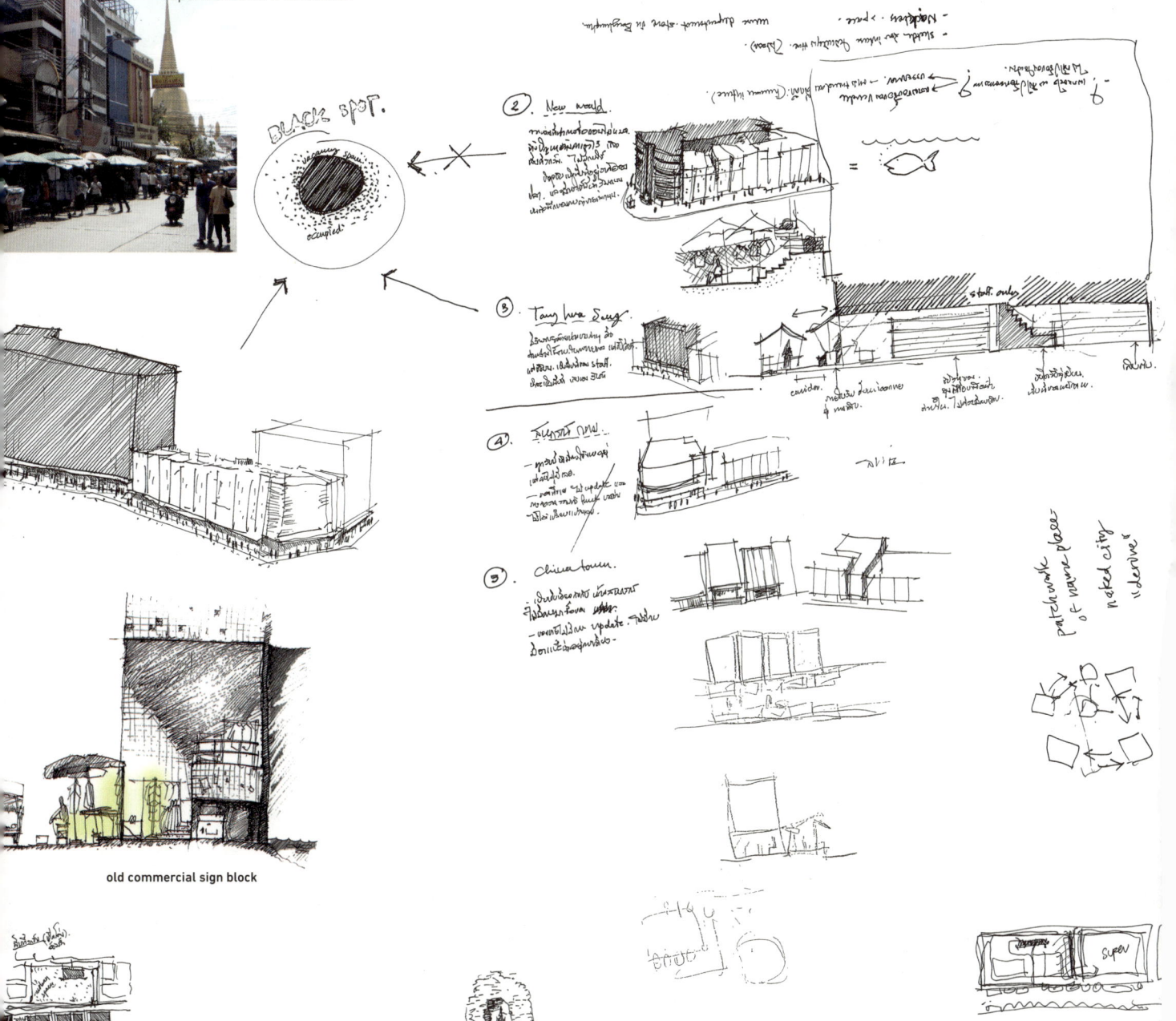

old commercial sign block

066　Dérive in BUNGLUMPHU

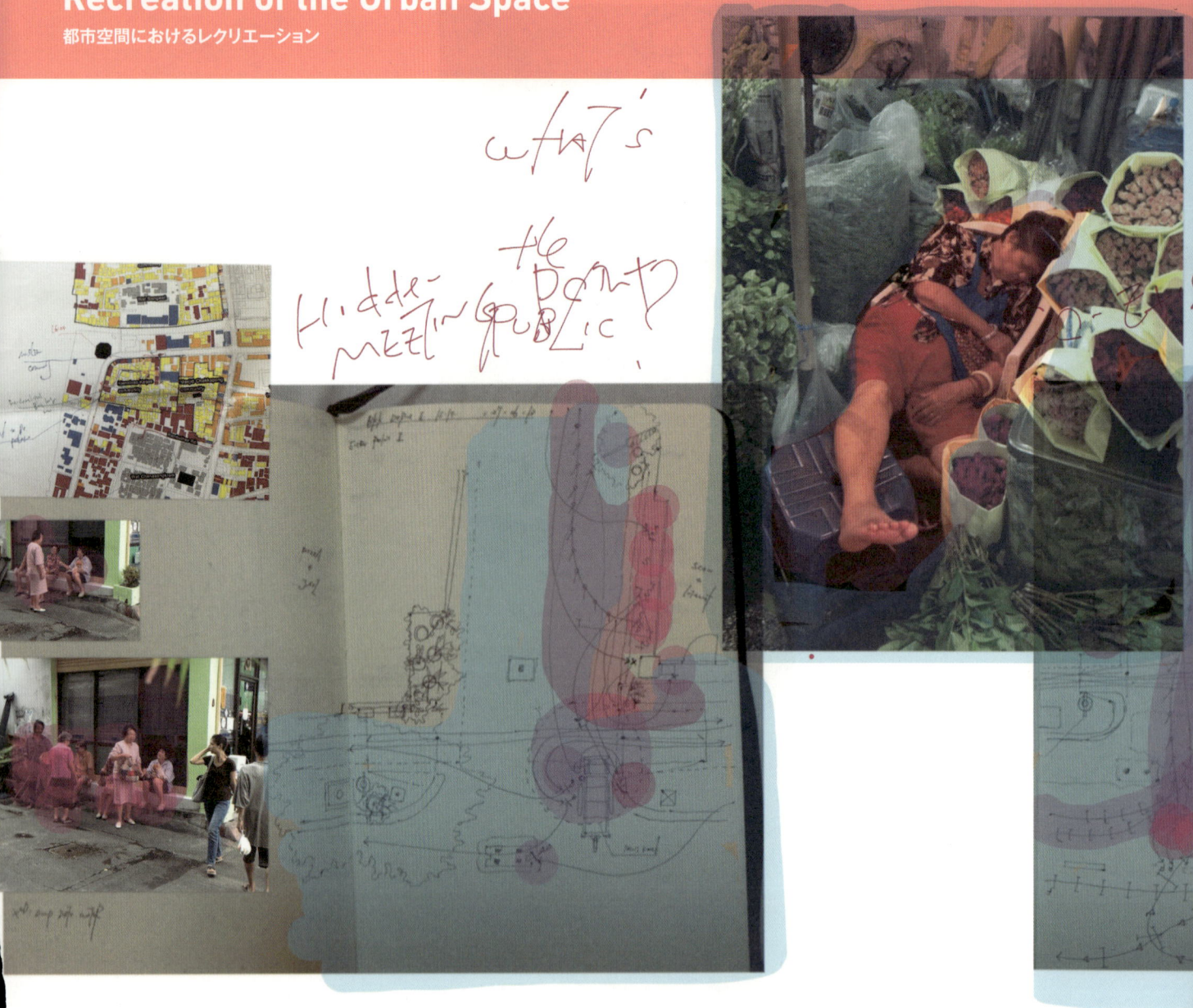

Recreation of the Urban Space
都市空間におけるレクリエーション

what's the hidden meaning public?

Dérive in_
Banglumphu
バンランプー

Drifter_
Tamao Hashimoto
橋本圭央

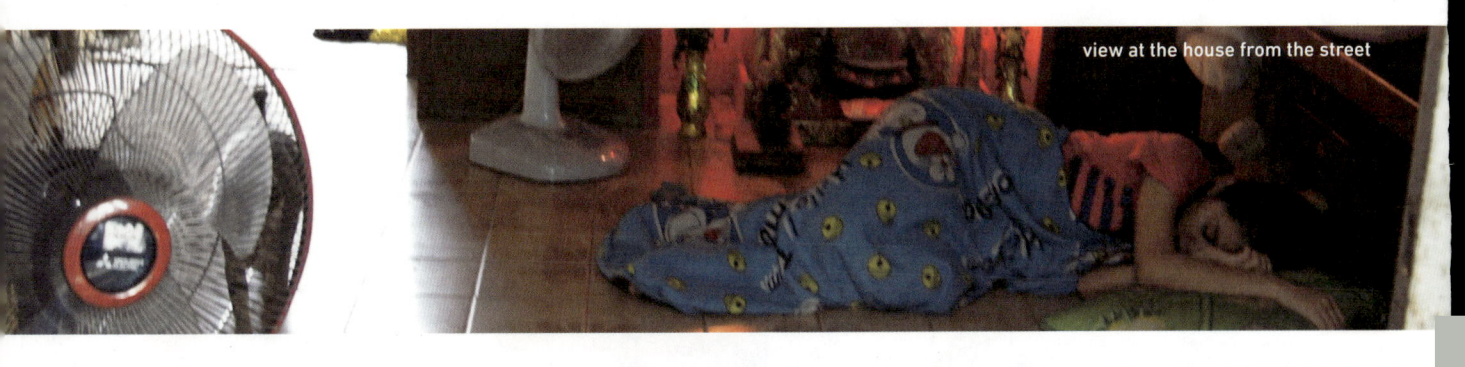

view at the house from the street

WELCOME
NEAT
FORMAL
CLEAN
INTIMACY
MESSY
DISCLOSE
WHOSE?
INFORMAL
MESSY
HIDDEN DOOR AMONG THE TREES
SURPRISED
BAD-SMELLING!
IT SEEMS MORE WIDER THAN IT WAS.
4
5
6
7
PRESSED!
LOST!
EXCITING!

070 | 071 | 072 Dérive in SUKHUMVIT

Exchange/Erode/Community
置換・浸食・コミュニティ

Dérive in_
Sukhumvit
スクンビット

Drifter_
Milica Muminović
ミリッツァ・ムミノヴィッチ

Trok Sok Soi
トロック、ソック、ソイ

Dérive in_
Sukhumvit
スクンビット

Drifter_
Thittayabhorn Mitudom and Sasinee Teangt
テッタヤーボーン・ミッツウドン、サシニー・テーンタワット

no exchange

diversity
no exchange
exchange
no exchange

main street

diversity
intensity
no exchange
exchange

diversity of communities-expressed in activities and architecture/but the life is always outside and architecture is always eroded by that life

eroding

From our discussion, we designed the dérive plan. Then we illustrated the assemblage of ten moments in "*trok*", "*sok*", and "*soi*" in Banglamphu.

the afternoon of 27 June, 2013

we walked through the narrow lanes *trok*, *sok*, and *soi*

K: Could *trok*, *sok*, and *soi*... which all refer to the narrow lanes... be defined simply based on their width?

C: I think... the widest lane is *soi*, then *trok*, and *sok*...

K: It's probably right... *soi* might refer to a wide lane that cars can pass through... however... it cannot be said that all *soi* are wide according to the perception or feeling of spaciousness... when we walked through Soi Samsen, I felt it was not wide... since we were aware of cars all the time... I felt more comfortable when we walked through some *trok*... even they were narrower than *soi*...

C: It seems that for definition of those spaces... the way we experience...the senses are more important...

K: I agree... I would like to add that the perception of environment might be shaped based on personal experiences of the place and the time when people get sensory information...

C: For that reason... we should do both ways... measure the width of the lanes... and to subjectively "measure" the experience...

Exchange/Erode/Community
置換・浸食・コミュニティ

Dérive in Banglumphu
バンランプー

Drifter Milica Muminović
ミリッツァ・ムミノヴィッチ

Dérive in BUNGLUMPHU 067 | 068 | 069

the life is always outside...it is exposed to the public...the boundary between public and private is pushed to the extreme...and it seems that in this point it gains a different meaning....that particular openness to the public becomes normal and somehow does not make me feel uncomfortable...

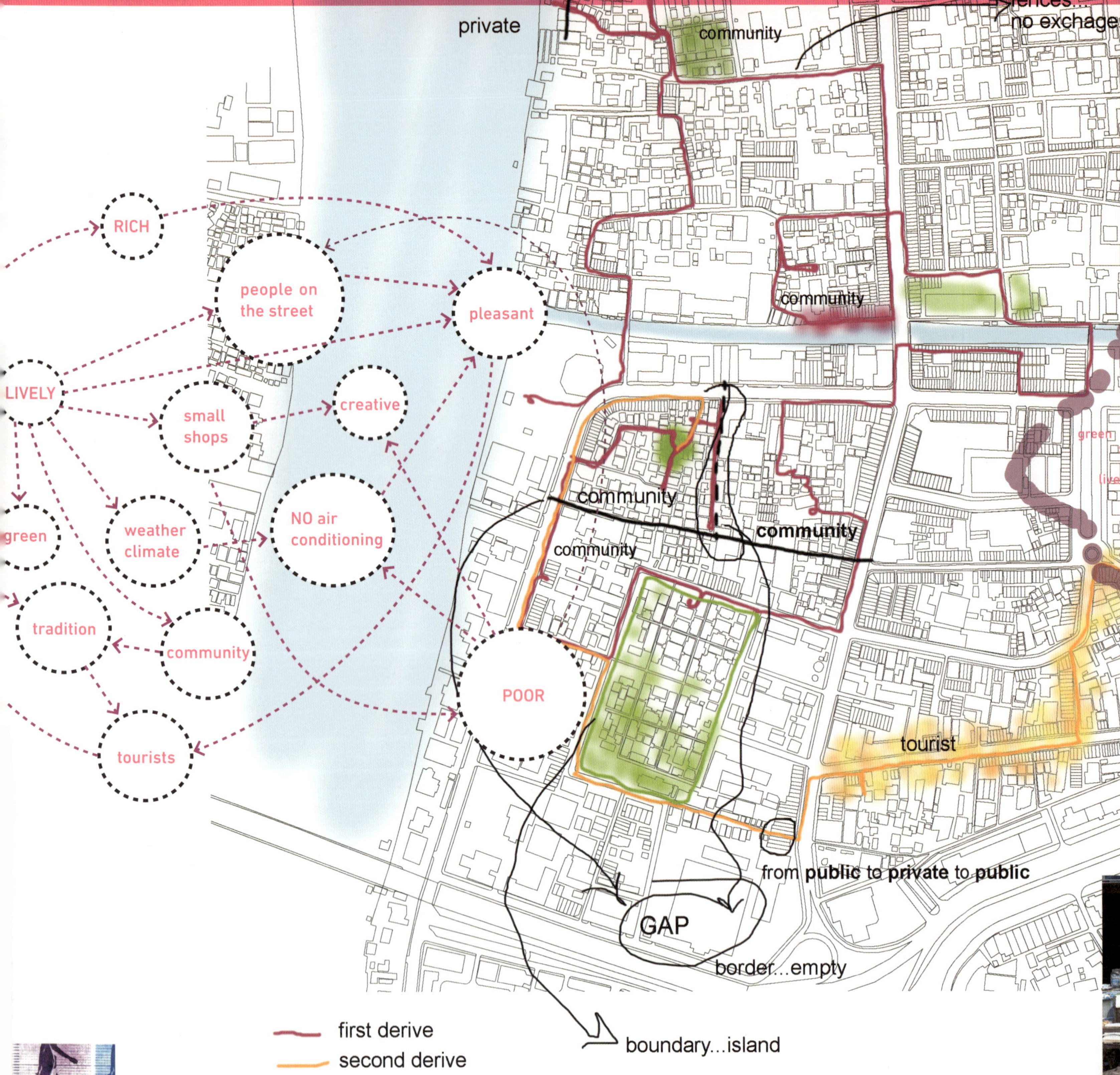

— first derive
— second derive

boundary
private
community
fences no exchange
community
community
community
community
community
from public to private to public
GAP
border...empty
boundary...island

RICH
people on the street
pleasant
LIVELY
small shops
creative
green
weather climate
NO air conditioning
tradition
community
POOR
tourists
tourist

public/private interface
intensity
private eroded
public green stretched lively
gap
intensity
difference

community
public stretched
private
intensity
private green
eroded
lively

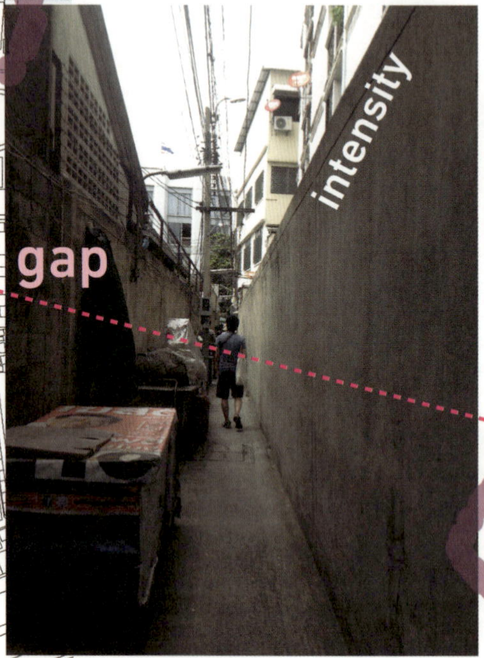

public/private interface inside (one side of the street) 
the same street (other side) public/private interface

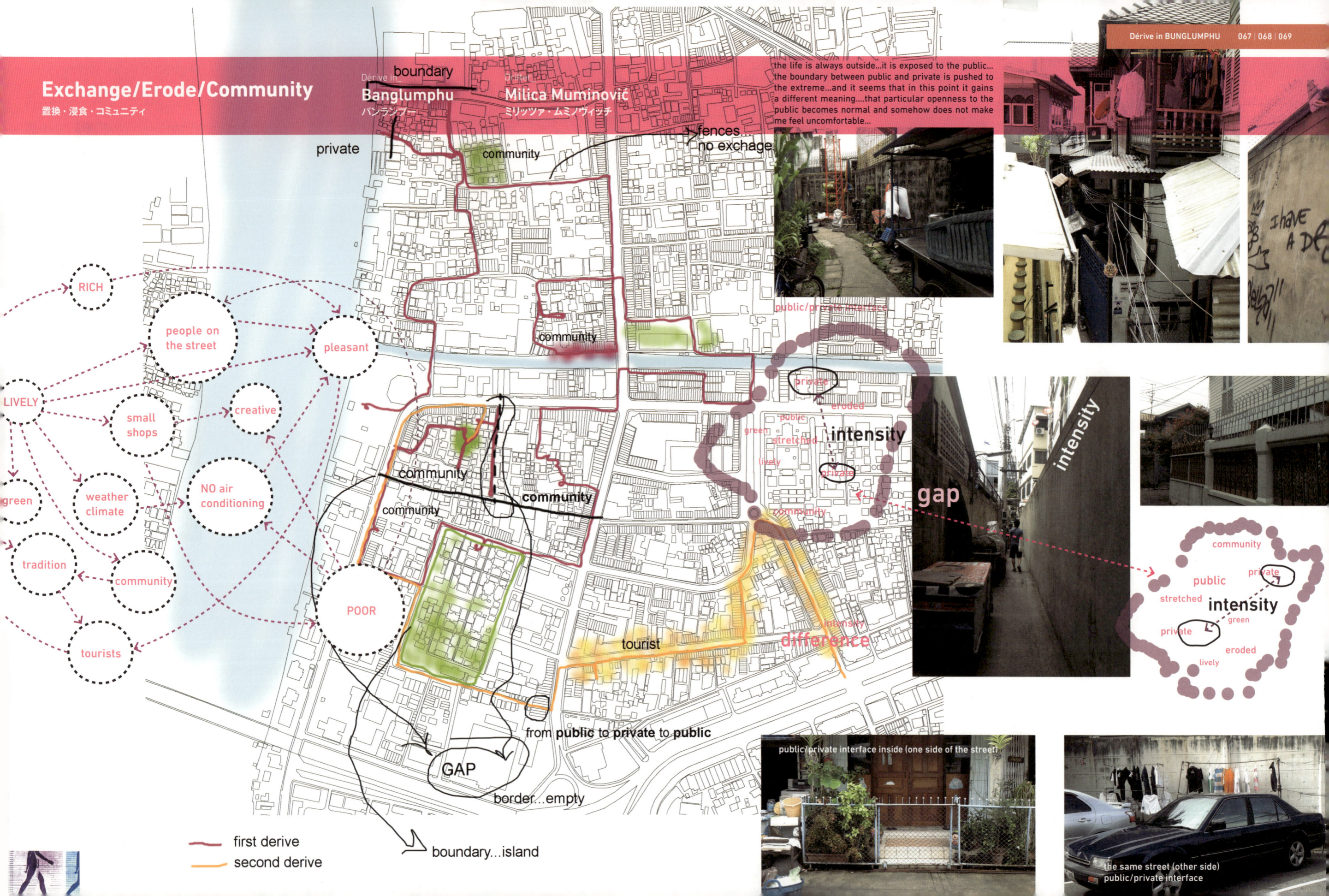

Dérive in SUKHUMVIT 073

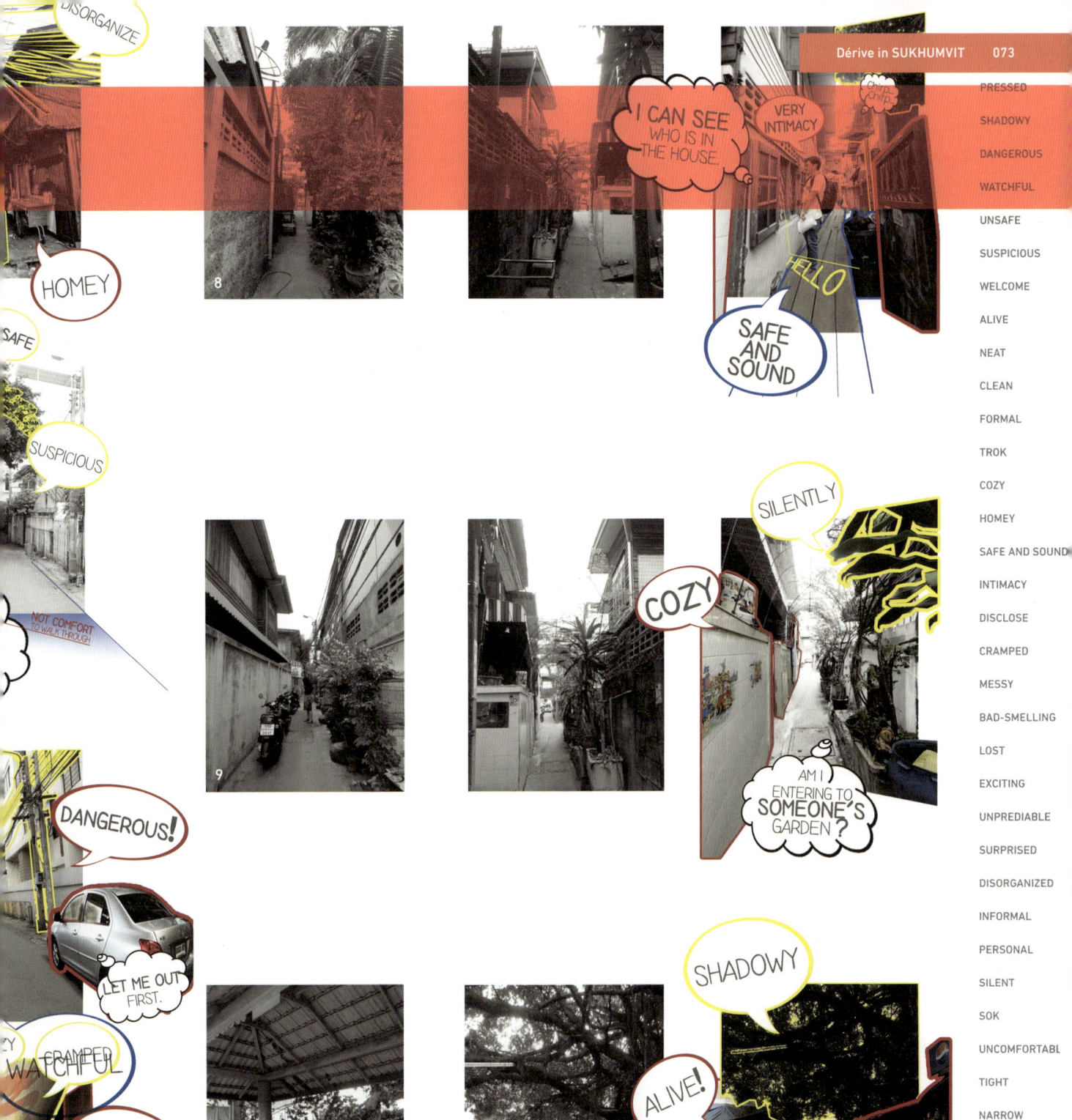

Fragments of Traditional Waterscape
伝統的な水辺の断片

Dérive in_
Sukhumvit
スクンビット

The first part: conventional waterscape

...was characterised through groups of wooden houses built on stilts packed along the narrow alleys.

At the end of the alley there was a wooden gangway. People used the gangway as a public-private interfacing space for meeting, hanging clothes, cooking and so on. There at the dead-end of the gangway was a house. The owner of the house allowed us to see the veranda nearby the river.

After that we walked through other small lanes. We saw groups of buildings which were lifted up over water and connected by wooden bridges. We found that almost all of the waterfront areas were privately owned, so they were used for only private activities...

My friends and I started the dérive in the afternoon of 3 July 2013...

...from the Phrasumane Fortress, we crossed the bridge and entered Wat Sangwet Community. We walked through the low land which used to be a swampy area outside the city wall.

...images of the passed times came to my mind... networks of waterways connected the area to Chao Phraya River and the other parts of Bangkok...then, waterways were the only kind of transportation... later, urban fabrics have been gradually altered canals have been filled and changed to streets... influenced by the development from European colonizes power since the 1850s.

coming back to the present...the area in front of us brings the sense of being close to the water... we decided to find the river...walking in zigzag towards the river...we found some traces of waterscape...based on their characters we could distinguish three different areas: 1. conventional waterscape; 2. water obstruction; 3. vulnerable area of new waterscape

The third part: vulnerable area of new waterscape

The last part we walked through a low-rise area located on the river bank. The community faced a severe flood during November 2011. This disaster reshaped the landscape. In the flood period, a new feature-one-meter up-lifting wooden walkway, was built and still remains here. Other features which were prepared for preventing flood were some sand bags packed along water bank and around a water drainage area.

When we dérived through this area, we could imagine how difficult it was for the people to live here during the flood period... and how they adapted their way of living to the new features of flood...

...we saw various kinds of persisting waterscapes which might be different from those in the past... although today the water is not well integrated in everyday lives...we found some well maintained waterways used for both public and private activities...

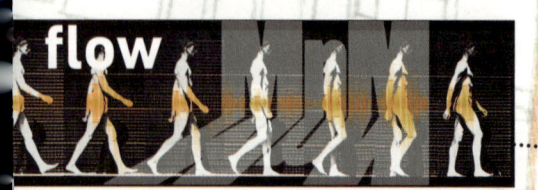

flow

Drifter_
Wimonrart Issarathumnoon
ウィモンモット・イッサラータッマヌーン

Dérive with
Davisi Boontharm and Thittayabhorn Mitudom

A gangway lifted up over water with its multi-uses

A remaining canal with its surrondings

A condominium obstructs the river front

The second part: water obstruction

The second part was characterised through large condominium obstructing the river front. On the opposite side of the condominium, we saw the remaining canal which was drained and blocked by the water gate. Unfortunately, as the canal has turned to drainage channel and become the back of the buildings, almost all of houses along the canal have no longer used the water front areas for daily activities. The guesthouse located on the corner of the canal was one of a few buildings which had open spaces connecting to the canal. The terrace on the second floor of the guesthouse is multifunctional space. People can enjoy the view to the waterway and get breezing wind.

New waterscape with one-meter uplifting walkway and sand bags

large forest-like green area in contrast with high-rise buildings in the city.

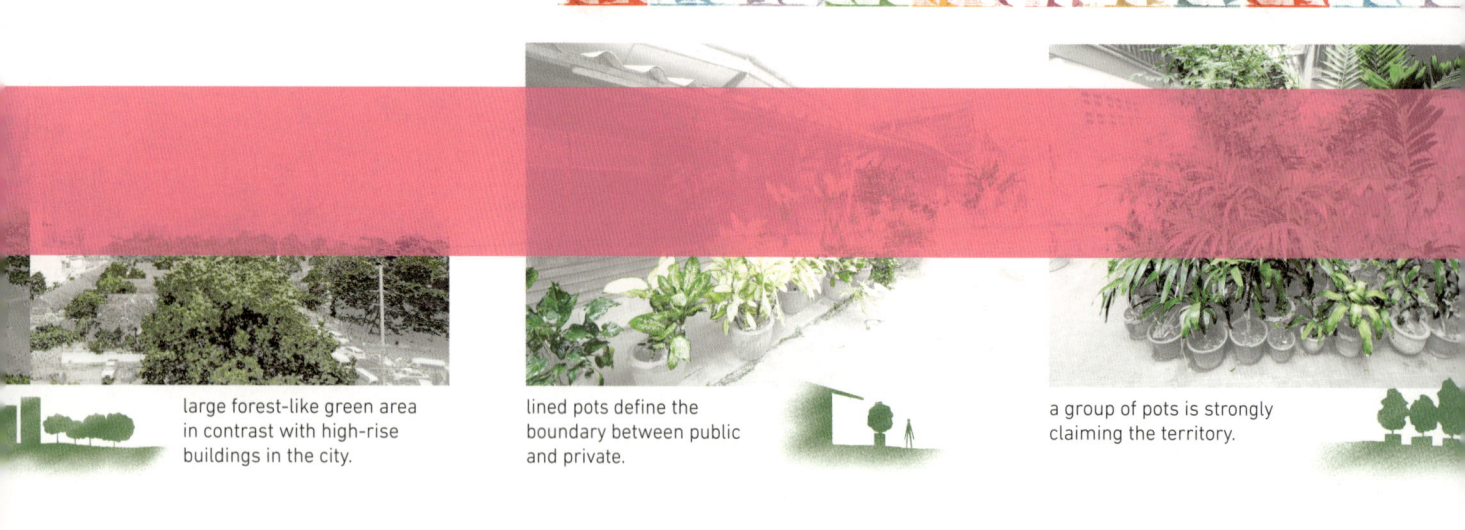

lined pots define the boundary between public and private.

a group of pots is strongly claiming the territory.

small market under large trees.

small greens are making sacred area by gathering at courtyard.

pots and other goods are occupying the street and it becomes part of the house.

A city is **floated** in greens. That was the first impression of the city, Bangkok. The cityscape is emerging from the **nature**, the architecture mingls with trees, and people put greens around their houses. The colors of nature strongly influence the **image** of the city with their vivid and colorful lives. Since it is tropical and hot, green has a critical role for the life of the people in the city....

...the green **connects** city and architecture, public and private, human and nature.

ops with canopies and trees ake a arcade-like street.

furniture and green creating outside living room.

small community space surrounded with green in a small street.

078 | 079 | 080 Dérive in SUKHUMVIT, BUNGLUMPHU

The fragmentation inside superblock
スーパーブロック内側の断片化

Dérive in_ Sukhumvit
スクンビット

Drifter_ Davisi Boontharm
ダヴィシー・ブンタム

Green สีเขียว 緑
Green สีเขียว 緑

Dérive in_ Banglumphu and Sukhumvit
バンランプー、スクンビット

Drifter_ Ken Akatsuka
赤塚 健

The idea was to make a short loop inside Sukhumvit superblock and go to the places which we already knew to visit a friend, see the bookshop, etc and try to capture the p/p interface along the way. The intention was not fully accomplished due to a minor injury of the teammate.

We made a dérive with moto-taxi through the main soi and short-cut soi. The walking dérive is in the middle of soi Ekkamai 12 and Thong Lor 10. We found a mixture of everything. There were full stream gentrified old buildings, the remnant of the old canal which remains like an oasis, the **juxtaposition** between cutting-edge **designs** of buildings or pockets of landscape and informal **everyday** survival activities like mobile vendors and moto-taxi stops.

The Sukhumvit soi is not made for pedestrian, just for private properties and cars, almost nobody walks in the soi.

City emerges in nature

houses are mixed with greens in the suburban area.

buildings and roads divided by trees.

Urban life happens under trees

a shop with tree becomes a community space for the residents.

movable shop under trees on the street.

Architecture mingles with greens

semi-outside space integrated in the small garden.

daily lives and trees are in very close relationship.

...à la recherche de rivière perdue

失われた川を求めて

Dérive in_ Banglumphu / バンランプー

Drifter_ Davisi Boontharm / ダヴィシー・ブンタム

Dérive in SUKHUMVIT, BUNGLUMPHU   075 | 076 | 077

Map annotations:
- CLOTH HANGING
- SMELL OF FRESH DETERGENT
- WOW! VIEW!
- WATER GATE
- WOODEN PLANKS
- BOAT FOR SALE 3 MILLION!
- SALA
- BLOCKED
- PUBLIC LAND FORBIDEN TO ENTER
- BLOCKED
- WHITE BOAT OF THE RESTAURANT
- "EAT WIND VIEW BRIDGE" RESTAURANT
- VIEW TO RAMA 8 BRIDGE
- WOODEN PLANKS
- BLOCKED
- GREAT VIEW OF RAMA 8 BRIDGE
- private resident - accessible by accident
- GIANT CONDOMINIUM
- public pier
- BOXING CLUB! ON DEAD-END STREET!
- GYM!
- SWIMMING POOL EMPTY
- XXXX WALL
- HUGE CAR PARK
- accessible under condition "pay to access"
- BURNT HOUSE FOR SALE
- GREEN STRIP
- GREEN STRIP
- GREEN WATER
- BLACK WATER CANAL SMELLY
- GREEN FIERCE DOG!
- CASUAL SITTING CHIT-CHAT GOZZIP
- WASHING MACHINES
- BIG BODI TREE
- TEMPLE & SCHOOL
- SCHOOL

This is not an aimless derive, We (Yui, her student and I) have a mission to get closer to the **riverfront** as much as possible. Along our journey are interesting encounters with the **local** people and fierce dogs. I became a **stranger** in my homeland. I found that the banality and authenticity of everyday life is **commodified** by the "guest house business". The everydayness became an asset for "experience economy".

The view of Chaophraya river is a sought after element for good business, suddenly the tight-knit village was punctured by a riverfront restaurant "swallow a breeze – view a bridge".

Dérive in BUNGLUMPHU 081

Where the wind flow
風の流れる場所

Dérive in_
Banglumphu
バンランプー

Drifter_
Kanyapat Seneewong Na Ayudhaya
カンヤパット・セニィウォン・ナ・アユタヤ

Dérive with_
Siwat Wittayalertpanya, Yoshiaki Kato and Sasinee Teangtawat

It was a warm afternoon of Bangkok on the 27th June 2013. I walked with my friends through Banglamphu's Trokkaijae, the old community closed to Santichai Prakan Public Park and Chao Phraya River.

Along the way we saw a very **narrow** alley which allowed only pedestrians to go through...It was not possible to serve even a small car and there were many things lying around. There were many activities in the area: a place for lunch, a place to park the strollers, a place for dries the clothes, etc. Looking up on the side, I found **dense** slum area. It gave me the feeling that every inch was important. On that day, there was no cloud and the sun was so bright. Along the street, there was a house without fence, the wall was adjacent to the walk way. It's windows and doors were widely **opened**. I could see people and activities inside, an old man, with no shirt on, was lying down watching TV on an artificial mattress with only one fan working.

I wondered how he could bear the hot without using any air conditioner?

community environment and how the wind flow

SG

Dérive in SINGAPORE ／シンガポール
Chinatown / Geylang / Little India / Tampines / Tiong Bahru

VISUAL ESSAY by **Ilze Paklone** and **Rafael A. Balboa**
イルゼ・パクロネ、ラファエル・バルボア

based on the work by: Balboa, Boontharm, Chia, Hashimoto, Lim, Muminović, Paklone, and Yeo
fieldwork by: co+labo Radović team and National University of Singapore, and Heng team

Singapore intensity – urban compression by Boris Kuk

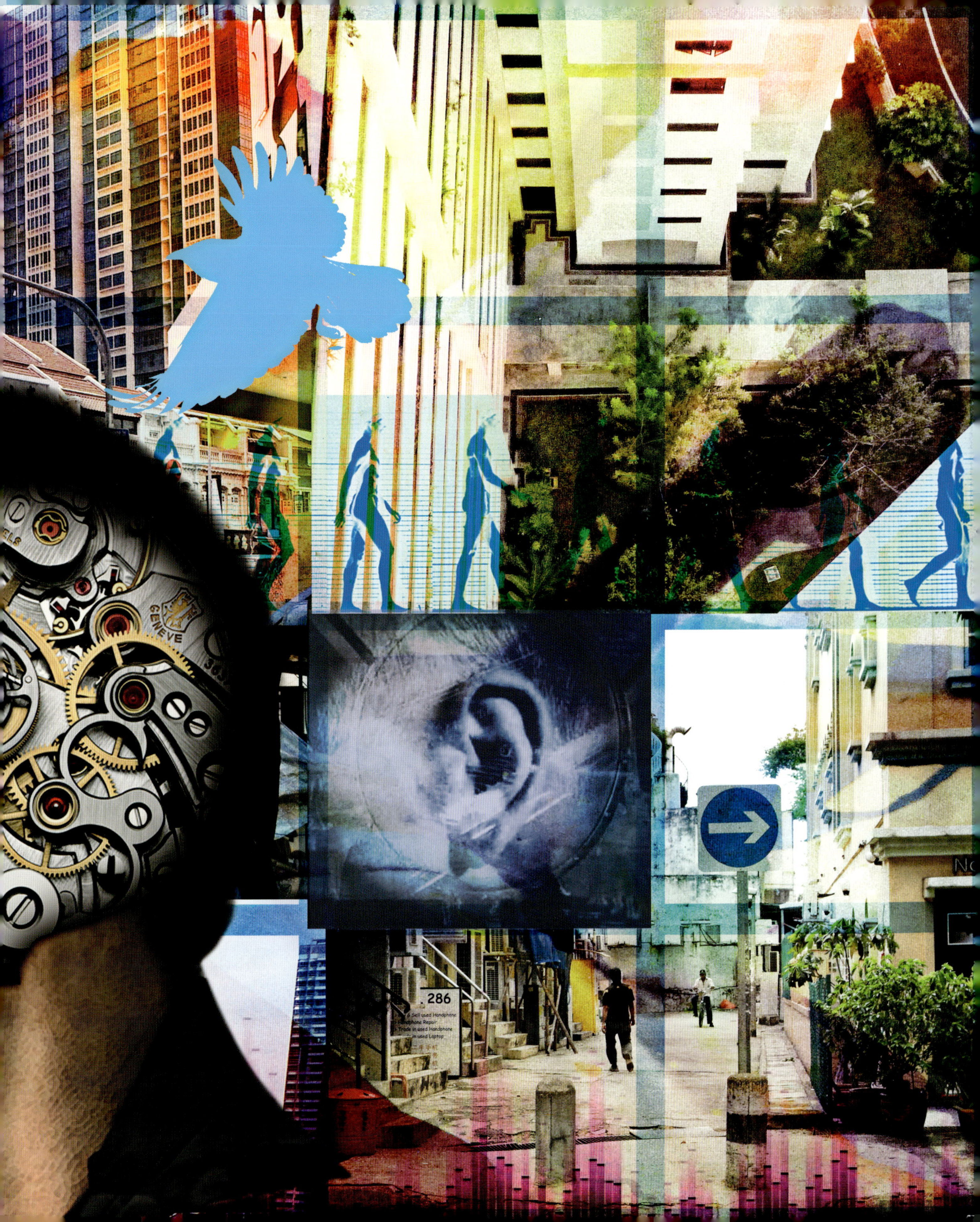
GENEVE
286

Drifter_
TAMAO HASHIMOTO
橋本圭央

Dérive in_
GEYLANG, LITTLE INDIA
ゲイラン、リトルインディア

In search for_
TRACING
BODY CHOREOGRAPHIES
身体的コレオグラフィーの透写

Key findings_
INTERACTIONS
FOSTERED BY HUMAN SCALE OF BUILDINGS

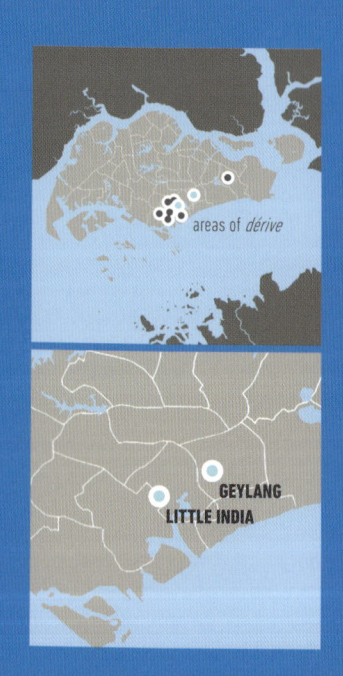

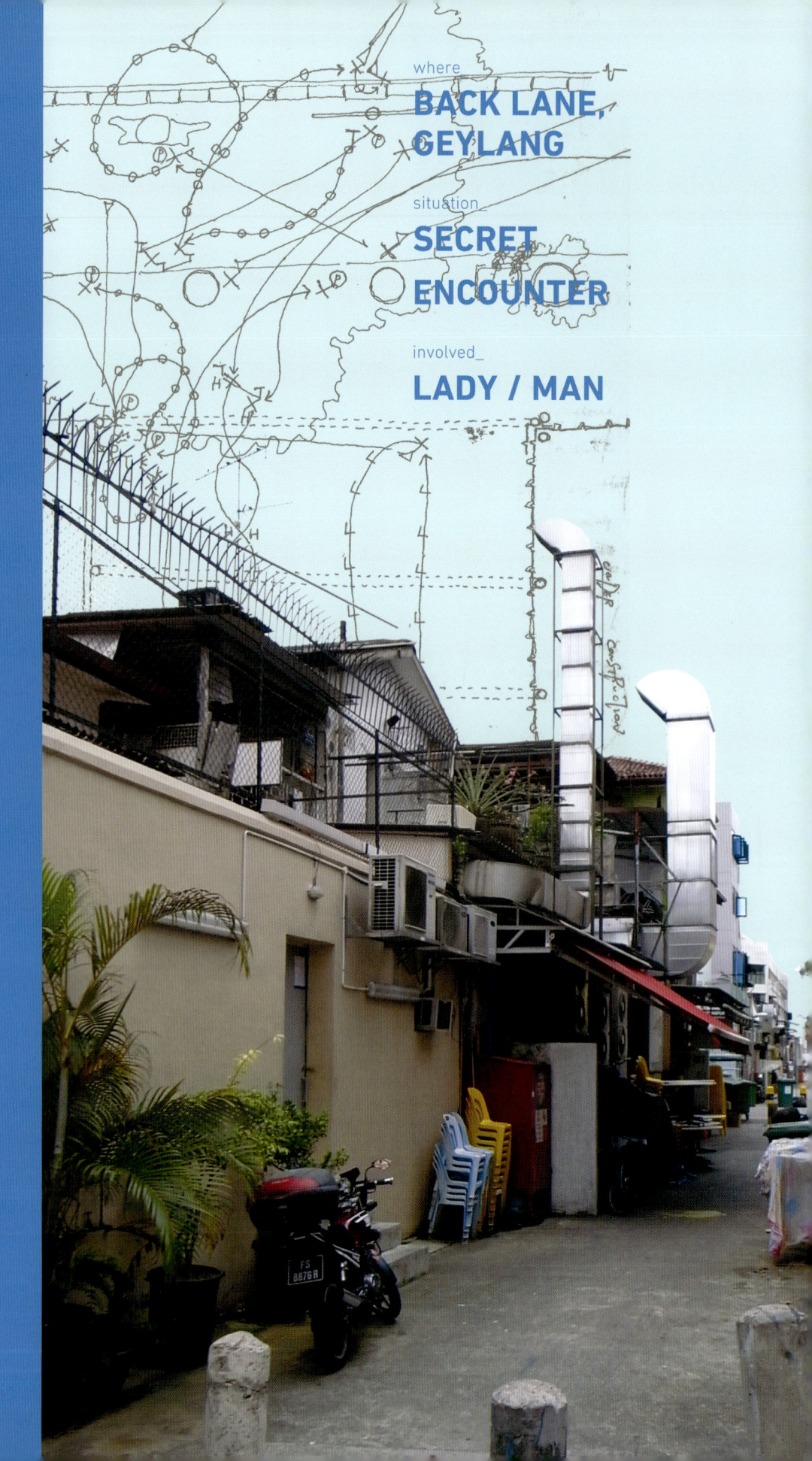

areas of *dérive*

GEYLANG
LITTLE INDIA

where
BACK LANE, GEYLANG

situation_
SECRET ENCOUNTER

involved_
LADY / MAN

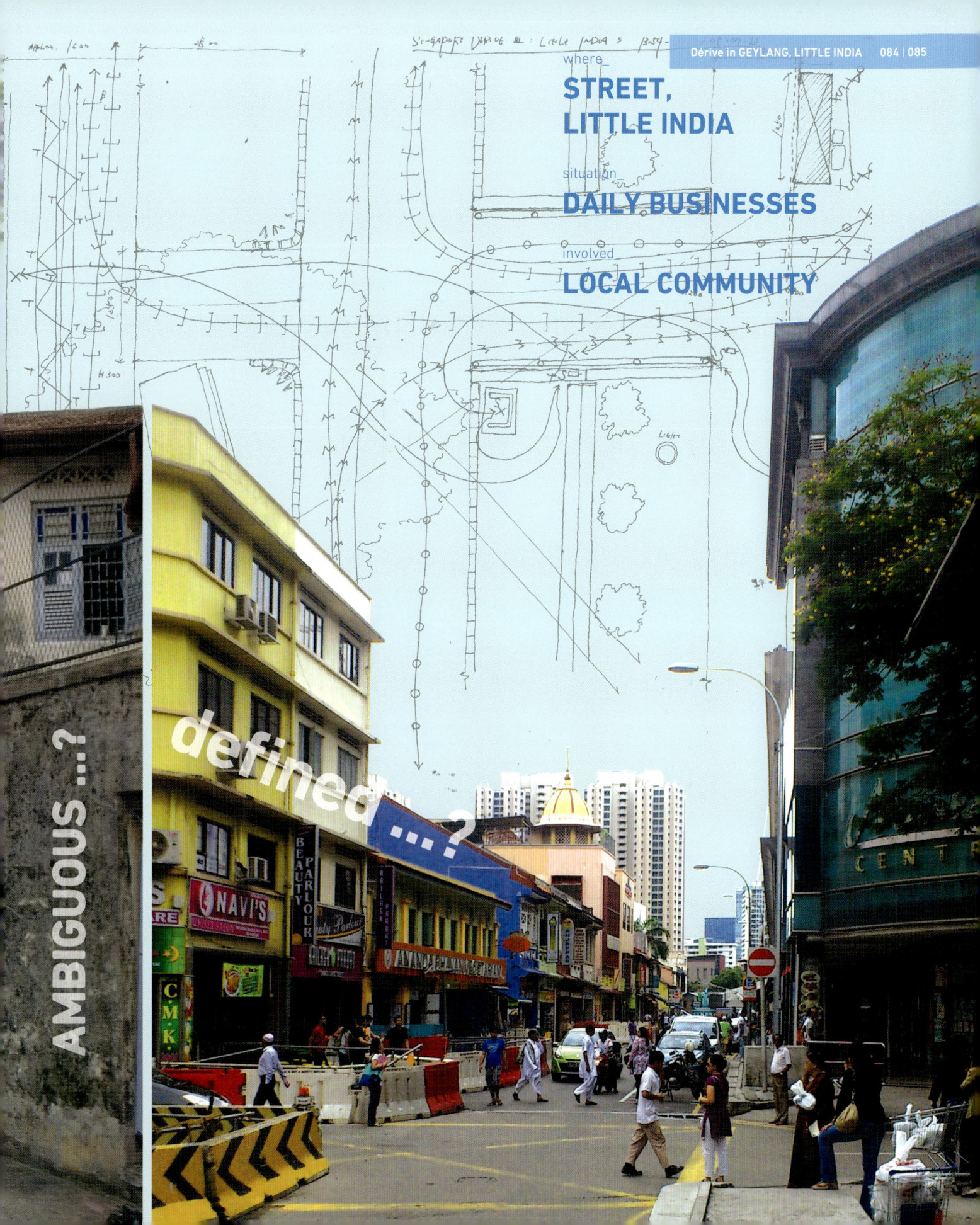

Dérive in GEYLANG, LITTLE INDIA   084 | 085

where_
STREET, LITTLE INDIA

situation_
DAILY BUSINESSES

involved_
LOCAL COMMUNITY

AMBIGUOUS ...?

defined ...?

Drifter_
MILICA MUMINOVIĆ
ミリッツァ・ムミノヴィッチ

Dérive in_
CHINATOWN, TIONG BAHRU, GEYLANG
チャイナタウン、チョンバル、ゲイラン

In search for_
DNA
TRACKING PROTOTYPES
DNA：原型の追跡

Key findings_
SHOPHOUSES
AS BASIS OF
URBAN LIFESTYLE

areas of dérive

CHINATOWN

↕ **intensity**

minimal
intense
history
crowded
no privacy
dirty

**private
hidden back**

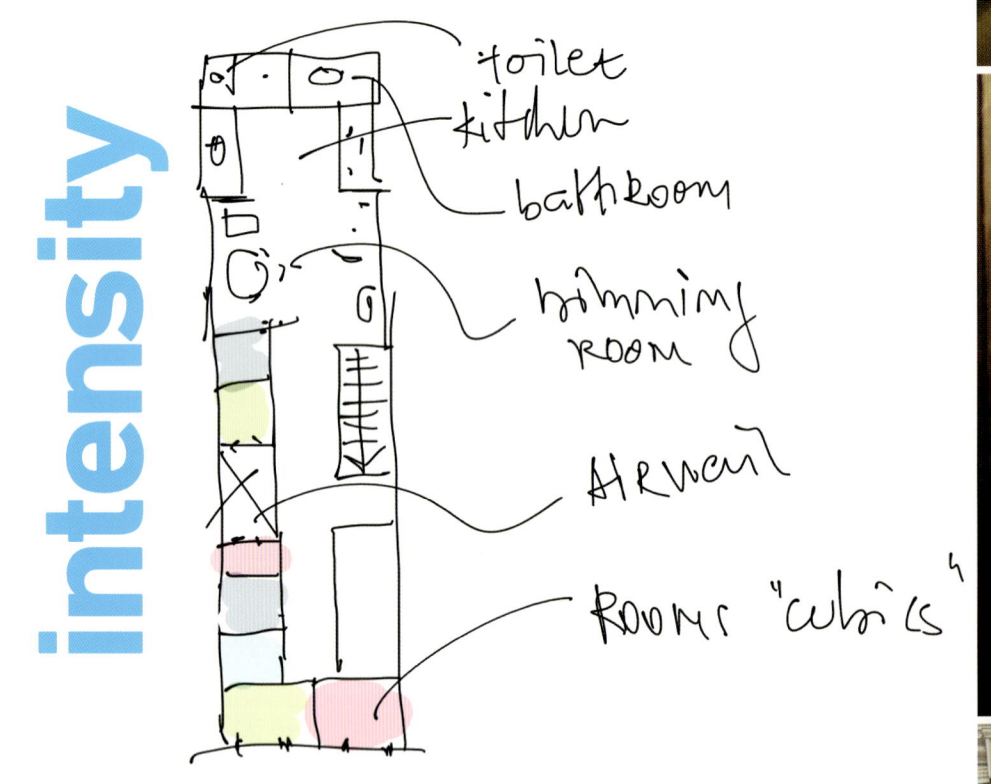

*toilet
kitchen
bathroom
dimming room
Airwell
Rooms "cubics"*

**public
exposed front**

semi-private semi-
PUBLIC
PUBLIC
celebrated
consumed
well-designed

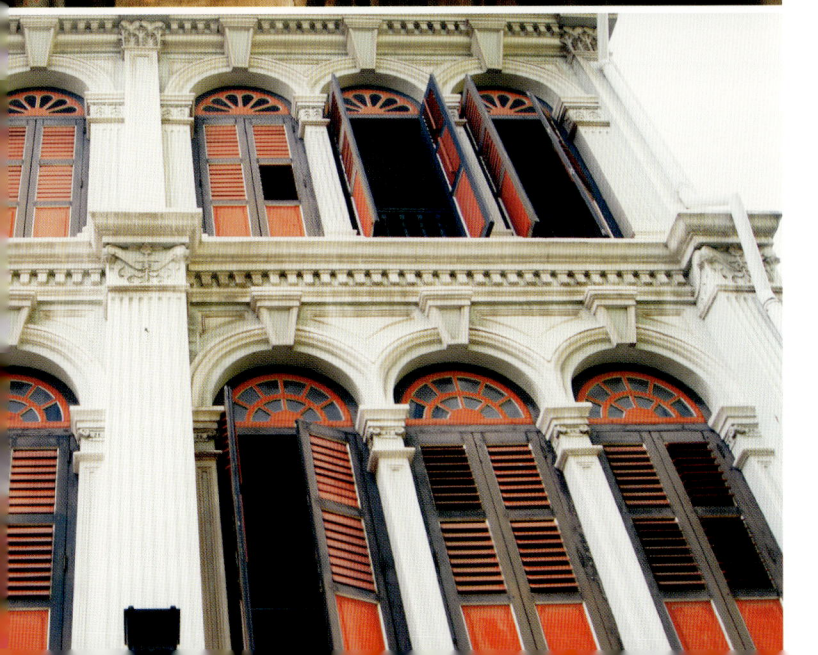

Dérive in CHINATOWN, TIONG BAHRU, GEYLANG   086 | 087

Behind the well designed, well organized, happily painted, incredibly maintained facades in streets of Chinatown...there was different image...hidden in history...the image of suffering...poverty...cramped spaces...dirty spaces...

The intensity appears in the impossibility to connect those...spaces...people...the architectural style...reminding me of Italy...and Chinese sense of size and cleanness...and although belongs to the past...it is there...celebrated and consumed....**the intensity is well designed...**

If we give our best to transport ourselves to the past...**we can imagine that real intensity between the hope of the new world and reality of today...or then...**

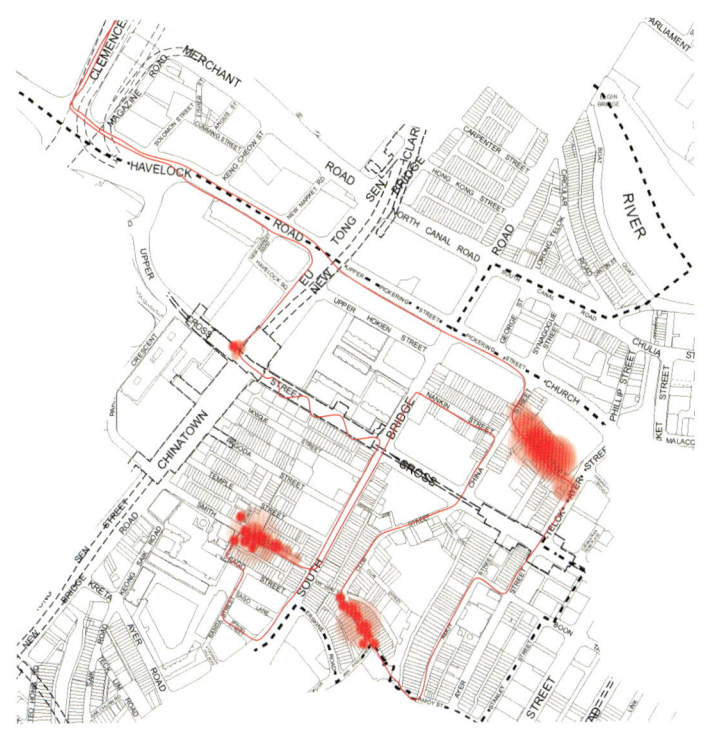

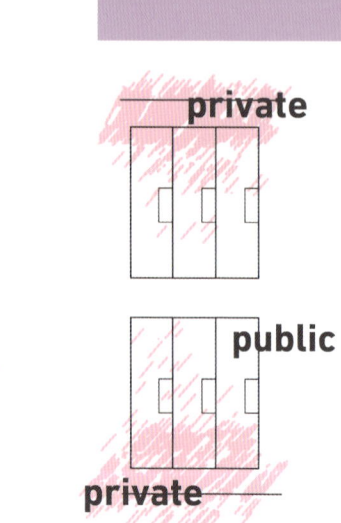

Key findings_
PRIVATE AND PUBLIC INTERFACE
AS EXTENSION OF PROTOTYPICAL ATTITUDE

CHINATOWN

Public   exposed   front   semi   private   public   communal   minimal   private   intense

EXPOSING THE PUBLIC

APPROPRIATING THE PUBLIC

HIDING THE PRIVATE

TIONG BAHRU

private

public

private

Dérive in CHINATOWN, TIONG BAHRU, GEYLANG   088 | 089

HIDING THE PRIVATE

EXTENDING OF THE PRIVATE

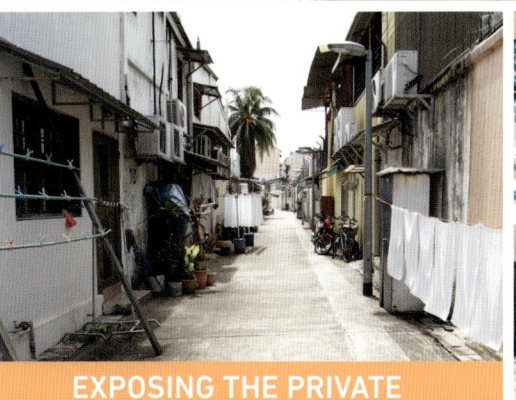

EXPOSING THE PRIVATE

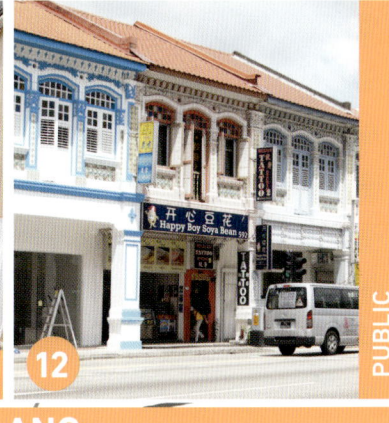

PUBLIC

GEYLANG

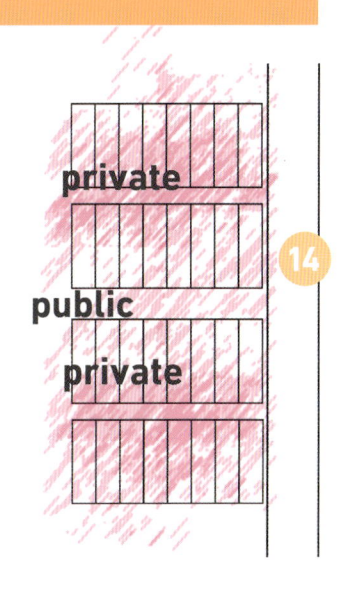

private

public

private

Drifter_
DAVISI BOONTHARM
ダヴィシー・ブンタム

Dérive in_
TIONG BAHRU
チョンバル

In search for_
SEMI LOCAL
LAYERS OF TRANSITION
半ローカル：変遷の層

Key findings_
DROPS OF INVENTIVE USE
GENTRIFICATION AT MICRO SCALE

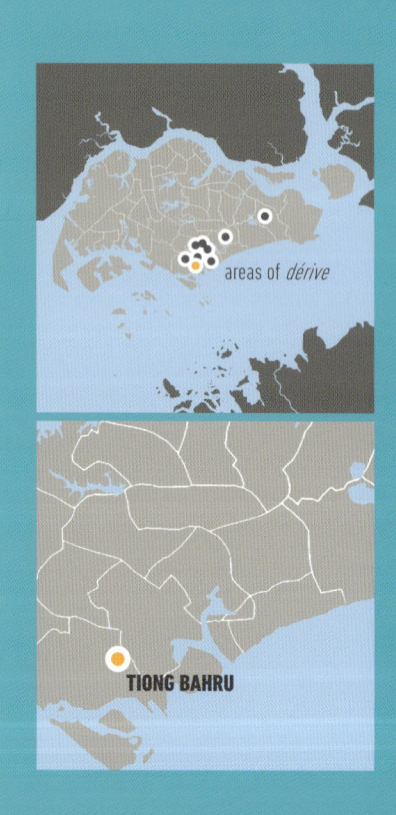

areas of *dérive*

TIONG BAHRU

Dérive in TIONG BAHRU 090 | 091

Drifter_
LIM SZEYING
リム・スゼイング

Dérive in_
TIONG BAHRU
チョンバル

In search for_
TRESPASSING
THRESHOLD OF PRIVATE
不当な侵入：
プライベートの閾値／限界値

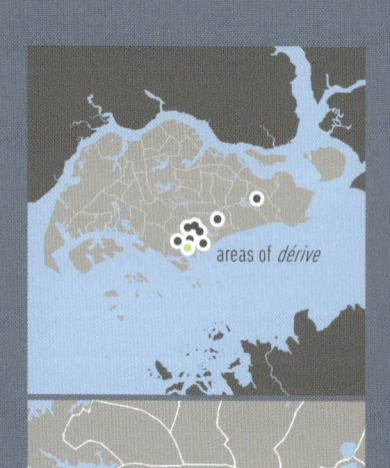

areas of *dérive*

TIONG BAHRU

THE HOUSING AND DEVELOPMENT BOARD (HDB) FLATS

HIGH-RISE RESIDENTIAL

COMMUNITY CENTER...?

Dérive in TIONG BAHRU 092 | 093

- GENEROUS GREENERY
- LOW RISE MIXED USE
- OPPORTUNITIES FOR OPPORTUNITIES?
- THE SINGAPORE IMPROVEMENT TRUST (SIT) FLATS
- LOW RISE RESIDENTIAL
- HIGH-RISE RESIDENTIAL
- MATURE AND TODAY'S HIPPIES…? CREATIVE?
- REJUVENATED ESTATES…AGING WITH STYLE…?

route of derive
open green space
low rise residential
low rise mixed use
typical shophouses
high-rise residental / mixed u

Key findings_
SIMILAR INDIVIDUALITIES
AT THE GROUND LEVEL AND HIGHER FLOORS

TIONG BAHRU

OPEN CORRIDOR, 40TH FLOOR HIGH-RISE RESIDENTIAL BLOCK

public ⟷ semi public ⟷ private

BACK BETWEEN TWO LOW RISE RESIDENTIAL BLOCKS

private ⟵ public ⟶ privat

Dérive in TIONG BAHRU 094 | 095

private items in the public realm

PUBLIC SPACE APPROPRIATIONS

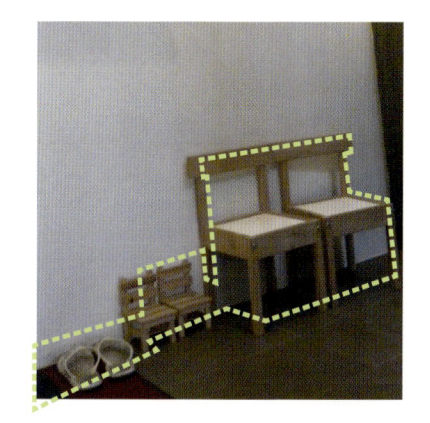

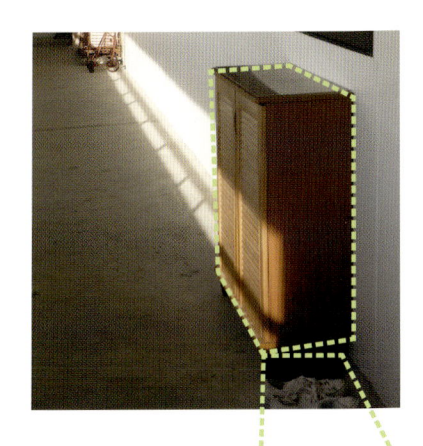

	LOW RISE RESIDENTIAL	LOW RISE MIXED USE AREA	HIGH-RISE RESIDENTIAL AREA
necessity	▬▬	▬	▬
convenience	▬▬▬	▬	▬▬▬
ordered	▬▬▬	▬▬	▬▬
restricting			▬▬
haphazard		▬	

Drifter_
LIM SZEYING
リム・スゼイング

Dérive in_
LITTLE INDIA
リトル・インディア

In search for_
ENCOUNTERS
AS MEASURE OF INTENSITY
遭遇：その強度を測定する

areas of *dérive*

LITTLE INDIA

Dérive in LITTLE INDIA   096 | 097

Farrer park Station

Adult-Industry / Desker Road Backlane

Express Medical Service / Blk 614

Rowell Road

Typical Indian Eatery / 8 Cuff Road

Local Market / Cambell Lane

Little India Community / Rowell Court

Drifter_
YEO SU-JAN
楊淑娟

Dérive in_
TAMPINES
タンピネス

In search for_
SENSING
HEARTLAND NARRATIVE
感知：中核地域の物語

Key findings_
FLUCTUATIONS
OF INTENSITY HIDDEN MICRO IMAGE OF THE LANDSCAPE

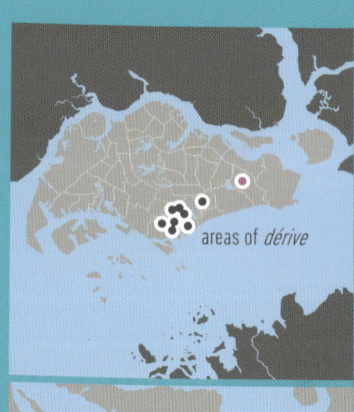

areas of *dérive*

TAMPINES

An ornate Chinese temple is a **contrast to the modern high-rise buildings** in the background, thus embedding its sacred urbane presence in the urban landscape.

04:39PM | SACRED

03:08PM | PLACID SPACE

The void deck evokes **a sense of 'time standing still'** which stands in stark contrast to the lively tempo of Tampines Central where commuters, shoppers, office workers, students, and the retirees converge to create a **heightened place rhythm.** Is the void deck 'under-utilized' space or a placid space?

Two residential **blocks, mirror-images** of each other, form a wall-like edge to a neighbourhood park;

Dérive in TAMPINES   098 | 099

A striking **wall mural** creates a sense of **surprise**

04:18PM | PLAYSCAPE

A commercial **fast food outlet** enjoys visibility with its street front location. Directly behind is a **'wet' market**

03:53PM | FRONT REGIONAL/BACK REGION

03:17PM | (DIS)SIMILARITY

03:33 | URBAN NICHE

03:42 | OBTRUSIVE DESIGN

A pocket garden tended by volunteers creates opportunities for the making and remaking urban public space.

The **monolithic architecture** of this residential block creates a concrete wall effect at the traffic junction; it is an **abrupt obtrusion** to the finer-grained, meandering footpaths within the housing estate.

Drifter_
PATRICIA CHIA
パトリシア・チィア

Dérive in_
TIONG BAHRU
チョンバル

In search for_
INTRUDING
FROM FACE TO REAR
侵入：
表裏への侵入

areas of *dérive*

TIONG BAHRU

i was mesmerized at how
between the

notions of 'face' and

encircling the block, i was captivated by

Dérive in TIONG BAHRU 100 | 101

the staircase had become an architectural element that play host to territorial overlaps neighbour upstairs and the neighbour downstairs.

'rear' cought me here. it seemed that all the life was out 'back' the smooth layers of brick. curious, i entered one of the shop units - it was a hipster restaurant

we stumbled into a **network of spinal pathways with long axial vistas**. spine, ramp, to back door, to back door again, spine again.

evening walkers? **GENEROUS GREENERY BUFFERED THE PUBLIC FROM THE PRIVATE**. neatly arranged at the junction of each private unit to the public path, sat a black dust bin.

SPINE

Key findings_
GENEROUS GREENERY
AS LAYER BETWEEN PRIVATE AND PUBLIC

TIONG BAHRU

Dérive in TIONG BAHRU 102 | 103

Drifter_
ILZE PAKLONE
イルゼ・パクロネ

Dérive in_
GEYLANG
ゲイラン

In search for_
NEGLECTED
OPPORTUNITIES OF BACKLANES
軽視：見過ごされた裏通りの機会

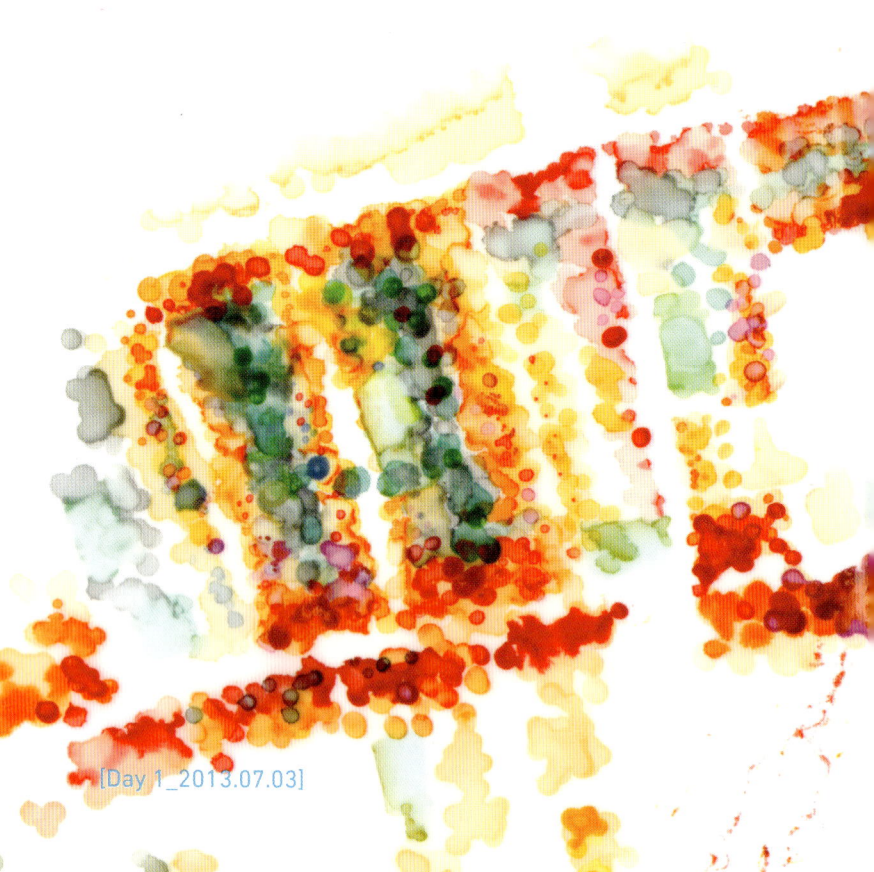

areas of *dérive*

GEYLANG

[Day 1_2013.07.03]

[08:00 – 11:45_ China Town_with Milica Muminović]

(in)tangible reminiscences ...how do transposed cultural practices become indigenous? ...when/how do continuously temporary moods accumulate into something enduring...what is that, 'something' that fades away, gets rooted, becomes an impenetrable set of concepts after metamorphoses have taken place...

[17:21 – 19:03_Tiong Bahru_with Szeying Lim]

still...the ground, plateau of communal practices stir ups of everyday cultural interactions...to loose anonymous residence at the 48th floor ...'occurring naturally in a particular way'...

Dérive in GEYLANG 104 | 105

[Day 2_2013.07.04]

[09:00 – 12:00_Central Business District_with Davisi Boontharm]

...city state as operations in
logistics...merchandize...controlled, regulated,
planned, created...authority
Richie describes **Edo/Tokyo as
'eternal present'** life where
one can renounce from daily routines...
Singapore?

[Day 3_2013.07.05]

[09:00 – 12:00_Geylang_with Davisi Boontharm and Patricia Chia]

thresholds...**regulated
and de-regulated**
practices...'measurements' of intensity...

backlanes...**services
mediation and
transients between**
programs and spaces, front and back streets,
private and communal...**softness of spaces...**
shared-ness

generic and
locally specific...appropriation of prototypes

Key findings_
REGULATED AND DE-REGULATED ACTIVITIES

Key findings_
DUALITY IN ORGANIZING THE CITY

Key findings_
DIVERSITY IN DE-REGULATED OCCURANCES

Key findings_
ORIGINALITY IN LOCAL CONNECTIVITIES

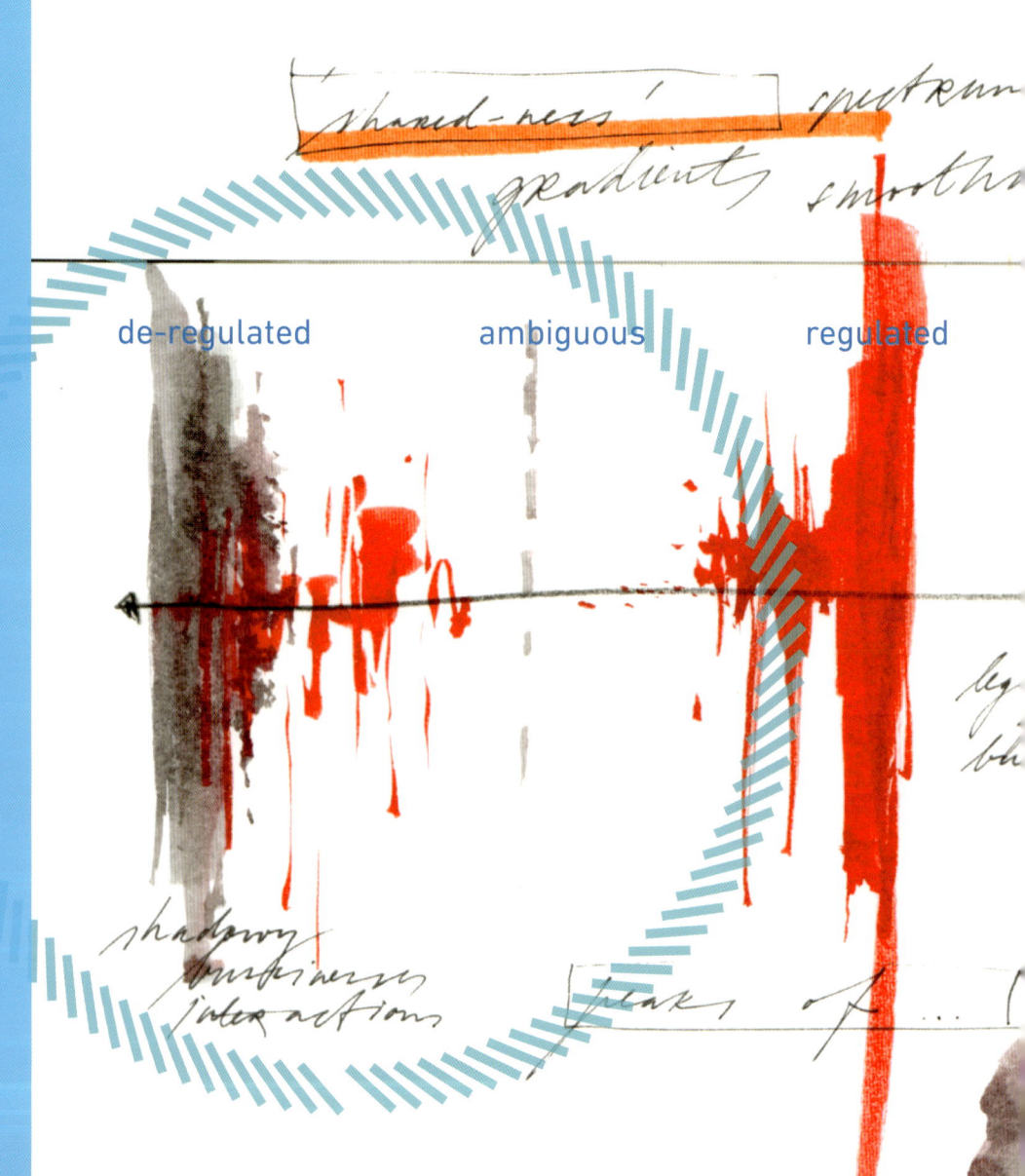

de-regulated   ambiguous   regulated

Dérive in GEYLANG 106 | 107

meeting your neighbour

ignorance

leaving households

some businesses

extracting shared spaces due to different reasons

de-regulated   ambiguous   regulated   de-regulated

public private/private interface

SHARE-
ABIL

because
one
occu

Key findings_
TYPICAL SHARED-NESS

AMBIGUOUS　　　REGULATED　　　PUBLIC | PRIVATE　　　AMBIGUOUS

Key findings_
IN-BETWEEN SPACE

BACK LANES AS POTENTIAL FOR DIVERSE COMMUNAL PROGRAMS

APPROPRIATING AMBIGUOUS SPACE OF BACK LANE

Dérive in GEYLANG

public (private/private interface

'CUT'

border

'5 fut' way

uniform, but very idealistic personal

| PUBLIC | REGULATED | DE-REGULATED | AMBIGUOUS | DE-REGULATED

'inbetween' gets very ambiguous
appropriated for variety of micro uses

INTENSE USE
OF BACK LANES !

Key findings_
SHARED-NESS OF THE AREA

Key findings_
FLUIDITY OF PROGRAMS

FULL/EMPTY CAFES
SHRINES
GATHERING SPACES
GREENERY
GROCERIES
SPORT CLUBS
GASOLINE STATIONS
OUTDOOR VENDORS
RESIDENTIAL
SHOPS
CLUBS

PUBLIC | PRIVATE   AMBIGUOUS   PRIVATE | PUB

REGULATED   DE-REGULATED   AMBIGUOUS   DE-REGULATED

MEASURING THE NON-MESURABLE
Jun 26-29, 2013
ASURING THE NON-MESURABLE
Jun 26-29, 2013
SURING THE NON-MESURABLE
ON-MESURABLE
Jun 26-29, 2013
BAN
-MESURA
May 24-26, 2013 DERIVE

THE EPILOGUE

エピローグ

Darko Radović | ダルコ・ラドヴィッチ

"'Every story is a travel story, a spatial practice,' writes de Certeau, and any theoretical system that tries to measure this story will inevitably exclude as much as it reveals. In this respect, de Certeau's theory of walking highlights the limitations of all systematic theoretical systems, psychogeography included, in accurately capturing the relationship between the city and the individual. ... Beneath the fabricating and universal writing of technology, opaque and stubborn places remain, claims de Certeau but, in our modern technological landscape, increasingly homogenous and regulated, dominated by surveillance and hostile to the pedestrian, it is now the novelist and the poet, not the theorist, who are uncovering and celebrating these overlooked and forgotten corners of the city" (Coverley, 2006).

That is because the novelist and the poet do not operate under restrictions which suffocate urban research and practice. *The urban needs to be liberated from the shackles of inappropriate practices which are abolishing some of the most essential qualities which make true cities – such as inclusion, equality, that very urban air which makes us free, the right to the city – in a word, urbanity.* They transform proud citizens into passive subjects of globalised consumerism.

In order not only to better understand *the urban*, but to help sustain the endangered, cardinal practices of *urbanity*, to continue being able to live and make cities, we need to create epistemological frameworks which would allow us to properly address their full complexity. In that process, urban research needs to embrace the study of whole systems, in non-reductive and methodologically inclusive ways. In a way comparable to the refrom undertaken in life sciences, that means inclusion of emergence, contingency, dynamic robustness and deep uncertainly and – more. Let's face it, urban complexity needs humility, well encapsulated in Franco Ferrarotti's approach to *the Other*: "I decide that I prefer not to understand, rather than to colour and imprison the object of analysis with conceptions that are, in the final analysis, preconceptions" (Dale, 1986; Radović, 2002).

A much-needed radical departure from current, aggressively imposed orthodoxies includes recognition of the importance and multiplication of subjectivities, and celebration of diversity and difference generated by concrete social and physical contexts of each investigated situation. Urbanism has to struggle against the reducionist, simplistic *solutionism* (Morozov, 2013).

Good urban research needs to be subjective objective, this way ...

References

Dale, R. 1986. *The Myth of Japanese Uniqueness*. Oxford: University of Oxford.

Morozov, E. 2013. *To Save Everything, Click Here*. New York: Public Affairs.

Radović, D. 2003. Celebrating the Difference. In *BMB Symposium*, eds. King, Panin, Parin. Bangkok: Kasetsart University Press, pp. 65–71.

Radović, D. 2008. The World City Hypothesis Revisited. In *World Cities and Urban Form*, eds. Jenks, Kozak, Takkanon. Abington: Routledge, pp. 121–136.

http://www.unhabitat.org/ (accessed 16.2.2014)

PARTICIPANTS

Hong Kong Team

Darko Radović
ダルコ・ラドヴィッチ

Davisi Boontharm
ダヴィシー・ブンタム

Milica Muminović
ミリッツァ・ムミノヴィッチ

Tamao Hashimoto
橋本圭央

Ken Akatsuka
赤塚 健

Hendrik Tieben
ヘンドリック・ティーベン

Thomas Chung
トーマス・チュン

Mika Savela
ミカ・サーベラ

Mo Kar Him
モー・カー・ヒム

Bill So
ビル・ソー

Daniel Ho
ダニエル・ホー

Cherry Wong
チェリー・ウォング

Della Cheung
デラ・チェング

Bangkok Team

Darko Radović
ダルコ・ラドヴィッチ

Davisi Boontharm
ダヴィシー・ブンタム

Milica Muminović
ミリッツァ・ムミノヴィッチ

Tamao Hashimoto
橋本圭央

Ken Akatsuka
赤塚 健

Yoshiaki Kato
加藤良章

Nuttinee Karnchanaporn
ナッティニ・カルンチャポン

Apiradee Kasemsook
アピラディ・カセンスーク

Wimonrart (Yui) Issarathumnoon
ウィモンラット・イッサラータッマヌーン

Siwat Wittayalertanya
シワット・ウィタカラータンヤ

Sasinee Teangtawat
サシニー・テーンタワット

Thittayabhorn Mitudom
テッタヤーボーン・ミッツウドン

Monsicha Kanittaprasert
モンシシャー・カンタープラサー

Thiti Kunajitpimol
ティッティ・クナッチピモン

Methi Laithavewat
メティ・ライタウィワット

Kanyapat Seneewong Na Ayudhaya
カンヤパット・セニィウォン・ナ・アユタヤ

Dherapat Sanguandekul
ティラパット・サンクンディクン

Prat Tinrat
プラット・ティナット

Parima Amnuaywattana
パリマ・アムアイワタナ

Singapore Team

Darko Radović
ダルコ・ラドヴィッチ

Davisi Boontharm
ダヴィシー・ブンタム

Milica Muminović
ミリッツァ・ムミノヴィッチ

Tamao Hashimoto
橋本圭央

Ilze Paklone
イルゼ・パクロネ

Heng Chye Kiang
王才強

Yeo Su-Jan
楊淑娟

Lim Szeying
リム・スゼイング

Patricia Chia
パトリシア・チィア

Apiradee Kasemsook | アピラディ・カセンスーク

Assistant professor at Silpakorn University, visiting lecturers at Chulalongkorn University, King Mongkut's Institute of Technology Ladkrabang, King Mongkut's University of Technology Thonburi, and many architectural institutions in Thailand and abroad. Api received her doctorate in Architecture from the Bartlett, UCL, and specialises in spatial morphology, particularly Space Syntax. Her main research interest is the spatial network and the way in which the network shapes or has been shaped by the configuration of buildings relating to the socioeconomic issue. Collaborated with Nuttinee Karnchanaporn for a number of projects, academically and professionally. They published a paper together, titled, 'World-class living,' in *World Cities* (2008, eds. by Jenck et al.). They were co-curators for Thailand Pavilion, for the 12th International Architecture Exhibition, Venice Biennale, 2010.

Darko Radović | ダルコ・ラドヴィッチ

Professor of Architecture and Urban Design at Keio University, heads the *Mn'M* Project. He is a co-director of International Keio Institute for Architecture and Urbanism – IKI and visiting Professor at the United Nations University, IAS in Yokohama. Darko's work focuses at the nexus between environmental and cultural sustainability. His books include *Green City* (2005, Routledge/UNSW Press; with Low, Gleeson, Green), *Urbophilia* (2007, University of Belgrade Public Art Public Space publishers), *Cross-Cultural Urban Design* (2007, Routledge, with Bull, Boontharm, Parin, Tapie), *Another Tokyo* (2008, University of Tokyo and ichii shobou), *Eco-Urbanity* (2009, Routledge), *The Split Case: Density, Intensity, Resilience* (2012, with Boontharm, Kuma and Grgić; flick studio), and *small Tokyo* (2012, with Boontharm; flick studio).

Davisi Boontharm | ダヴィシー・ブンタム

Architect and urbanist. She has lived and worked in Paris, Bangkok, Tokyo, Singapore and Melbourne. Her teaching and research interests include culturally sustainable architecture and urbanism, commercial space and creative milieu in Asian Cities (Tokyo, Bangkok, Singapore), Southeast Asian vernacular urban form (shophouse). She is currently Project Associate Professor at Leading Graduate School, Keio University, Japan. Her recent research books include *In the Search of Urban Quality: hundred maps of Kuhonbutsukawa street* (IKI and flick studio, 2014), *Tokyo, Bangkok, Singapore: Intensities, Reuse and Creative Milieu* (IKI and flick studio, 2013), *Future Asian Space* (NUS Press, with Hee and Viray, 2012), *small Tokyo* (IKI and flick studio with Radović, 2012). Her interest in cities also found its expression in creative work. She has exhibited drawings and paintings in Tokyo, Split and Vis (Croatia).

Heng Chye Kiang | 王才強

Professor and Dean of the School of Design and Environment, National University of Singapore. Teaches architecture and urban design and has lectured widely in Europe and Asia. He serves as member of several editorial boards of international journals and as jury member of many international design competitions in Asia. He is also Board member of the Jurong Town Corporation and Centre for Liveable Cities, Singapore. His research covers the urban design and history of Chinese cities. He consults internationally and is the conceptual designer of several international urban design/ planning competition winning entries in China. He publishes widely on urban history and design. His books include *Cities of Aristocrats and Bureaucrats* (1999), *A Digital Reconstruction of Tang Chang'an* (2006), and *On Asian Streets and Public Space* (2010).

Hendrik Tieben | ヘンドリック・ティーベン

Architect, urban designer, and an Assistant Professor at the School of Architecture of the Chinese University of Hong Kong (CUHK). He received his architectural education in Germany, Italy and Switzerland and holds a doctoral degree from the Swiss Federal Institute of Technology (ETH Zurich). Before arriving in Hong Kong, Hendrik Tieben has taught at ETH Zurich, HTW Chur, and CIA (Chur). At CUHK, he teaches urban design and theory and coordinates the Master Thesis and the Concentration Area Urban Design.

Ilze Paklone | イルゼ・パクロネ

She is an architect holding Professional Diploma and Master Degree in Arch. from the Riga Technical University (Latvia). She has worked in architecture offices in Riga, most notably NRJA and extended her professional experience internationally as junior architect at Wiel Arets Architects in The Netherlands with the Leonardo da Vinci Lifelong Learning Programme scholarship. Currently she is a Ph.D. student in Prof. Yukio Nishimura Urban Design and Conservation Laboratory at the University of Tokyo. Her current research is focused on mapping visually relationship between complex urban patterns and urban planning legal requirements. She is a contributor based in Japan for Domus magazine (Italy).

Nuttinee Karnchanaporn | ナッティニ・カルンチャポン

Lecturer in Interior Architecture Program at School of Architecture and Design, King Mongkut's University of Technology Thonburi, Thailand. Trained as an interior architect, she is working in various fields: teaching, research, interior architectural design, and writing on design related issues. She obtained her PhD in History and Theory of Architecture from Architectural Association. In recent years, Nuttinee has conducted researches that take interests in the way domestic life is engaged or ignored through the space of home in Bangkok. With the shared interests on 'urban domesticity,' she collaborates with Apiradee Kasemsook on the Mn'M research project. Current publications include *Rethinking Bangkok Domesticity: A Dialogue between film and critical design thinking* (2012), How small is too small? Bangkok (frugal) living (2011).

Milica Muminović | ミリッツァ・ムミノヴィッチ

Assistant Professor at University of Canberra, Faculty of Arts and Design. She holds PhD from Keio University. Milica has held the position of Visiting Junior Research Fellow at the Faculty of Science and Technology, Keio University, Japan, of teaching assistant at Faculty of Architecture, University of Novi Pazar, Serbia, of research assistant at Global COE Program and of teaching assistant at Keio University in Japan. Her present research extends the professional experience through studies conducted in Japan about identity, places, spaces in between architecture and urban design, public and private, with emphasis on residential architecture in Tokyo.

Tamao Hashimoto | 橋本圭央

He completed his BFA degree at Tokyo University of the Arts. After he worked as a freelance designer for several years in Japan, he entered the Architectural Association and received the AA Intermediate and AA Diploma. His drawings were featured in the publication Scales and/of Engagement. Since 2010, he has been teaching at Tokyo University of the Arts, while engaging in a series of international workshops with The Queensland University, Vienna Art Academy, Hong Kong Chinese University, the Architectural Association, and exhibitions. He is currently conducting a study on notation of the city.

Thomas Chung | トーマス・チュン

Associate Professor at the School of Architecture, The Chinese University of Hong Kong. He read architecture at the University of Cambridge, and has practiced as a registered architect at the United Kingdom. His research interest involved understanding how architecture contributes to the urban order of the modernity with respect to the broader cultural ground in question. He is currently researching on the interplay of architecture with urban representation and cultural imagination in Hong Kong, and the spatio cultural metabolisms of urban vernacular in the Asian city. Besides research, he is also actively engaged in interpreting theoretical discourse through exhibitions and creative works. He has curated or exhibited in every Hong Kong & Shenzhen Bi-City Biennale of Urbanism/Architecture since 2007, as well as Venice Architecture Biennale in 2010. His recent book publication includes *Refabricating City: A Reflection* (2010, Oxford University Press)

Wimonrart (Yui) Issarathumnoon | ウィモンラット・イッサラータッマヌーン

Assistant Professor at Department of Architecture, Faculty of Architecture, Chulalongkorn University, Thailand. She obtained her PhD degree in Department of Urban Engineering, Graduate School of Engineering, The University of Tokyo in 2009. Her research focuses on Architectural and Urban Conservation. Her publications include "Cultural Heritage Place: Yanaka, Tokyo and Banglamphu, Bangkok," in *Another Tokyo* (2008, The University of Tokyo and ichii shobou), "The Signs of the Rebirth of Japanese Upland Village Cultural Landscape," in *Najua* (2011, Journal of Faculty of Architecture Silpakorn University), and *Meanings and Principles of Urban Heritage Conservation: A Case Study of Yarn Banglamphu, Krung Rattanakosin* (2013, Chulalongkorn University).

Yeo Su-Jan | 楊淑娟

Ph.D. candidate in the Department of Architecture, School of Design and Environment, National University of Singapore. Her doctoral research examines the implications of contemporary urban nightlife in the design of public spaces and planning of local neighbourhoods. Su-Jan graduated from Simon Fraser University with a Bachelor of Arts in Geography (2003), followed by a Masters in Urban Development and Design from The University of New South Wales (2005). Prior to her PhD candidacy, Su-Jan worked as an urban planner at the Urban Redevelopment Authority in Singapore.

Subjectivities in Investigation of *the Urban*: The Scream, the Shadow and the Mirror

First Published by	IKI (International Keio Institute)+flick studio co., ltd.
Supervised by	Darko Radović / IKI (International Keio Institute for Architecture and Urbanism)
Author	Darko Radović
Graphical Design by	Authors of visual essays Milica Muminović, Ilze Paklone, Tamao Hashimoto and Rafael A. Balboa, based on the material from Keio University co+labo radović team, Chinese University of Hong Kong teams, Chulalonkkorn University team, Silpakorn University team and National University of Singapore team. Boris Kuk, author of the opening pages / compressions which introduce each of the three essays, and threads of visual identity throughout those essays
Technical Editor	Shinya Takagi and Takako Ishida (flick studio)
Technical Coordination	Tamao Hashimoto and Heide Imai
Book Designed by	Tomofumi Yoshida (9P), Yu Nakao
Production Coordinator	Kumi Aizawa (silent voice)
Japanese Translation	Tac Co., Ltd.
Published in Japan by	flick studio / Shinya Takagi
	2-28-6 Higashiazabu Minato-ku, Tokyo Japan 106-0044
	tel : + 81 (0)3-6229-1501 / fax : 03-6229-1502
Printed in Japan by	Fujiwara Printing Co., Ltd.

Copyright for publication ©2014 IKI – International Keio Institute for Architecture and Urbanism, Darko Radović and flick studio co., ltd. /
Copyright for photos and drawings, unless stated otherwise, is held by individual authors.
All rights reserved / Copyright@2014 for first edition of the book by Keio University

IKI / International Keio Institute for Architecture and Urbanism co+labo, Keio University, Department of Systems Design Engineering,
3-14-1 Hiyoshi, Kohoku-ku, Yokohama, 223-8522, Japan, tel: +81 (0)45-566-1675